Lecture Notes in Mathematics

Edited by A. Dold and B. Eckmann

1208

Sten Kaijser
Joan Wick Pelletier

Interpolation Functors and Duality

Springer-Verlag
Berlin Heidelberg New York London Paris Tokyo

Authors

Sten Kaijser
Uppsala University, Department of Mathematics
Thunbergsvägen 3, S-752 38 Uppsala, Sweden

Joan Wick Pelletier
York University, Department of Mathematics
4700 Keele Street, North York, Ontario, Canada, M3J 1P3

Mathematics Subject Classification (1980): 46M15, 46M35

ISBN 3-540-16790-0 Springer-Verlag Berlin Heidelberg New York
ISBN 0-387-16790-0 Springer-Verlag New York Berlin Heidelberg

Library of Congress Cataloging-in-Publication Data. Kaijser, Sten. Interpolation functors and duality. (Lecture notes in mathematics; 1208) Bibliography: p. Includes index. 1. Linear topological spaces. 2. Functor theory. I. Pelletier, Joan Wick, 1942-. II. Title. III. Series: Lecture notes in mathematics (Springer-Verlag); 1208.
QA3.L28 no. 1208 510 s 86-20242 [QA322] [515.7'3]
ISBN 0-387-16790-0 (U.S.)

Printing and binding: Druckhaus Beltz, Hemsbach/Bergstr.
2146/3140-543210

CONTENTS

CHAPTER 0

INTRODUCTION

Duality is one of the most important notions of functional analysis (and of modern Mathematics in general). It is thus not surprising that in the theory of interpolation spaces much attention has been devoted to duality questions. There are, however, some intrinsic obstacles that prevent the formulation of a good duality theory in the present setting of interpolation theory, which is the category of Banach couples. The first difficulty is that if $\overline{X} = (X_0, X_1)$ is a Banach couple, then the dual spaces (X_0', X_1') need not be a "dual couple". A necessary condition for the dual spaces to form a Banach couple is that $\overline{X}$ be what is usually called a "regular" couple, meaning that the intersection, $\Delta\overline{X} = X_0 \cap X_1$, is dense in both X_0 and X_1. However, as examples show, even if one restricts attention to regular couples, it turns out that the dual couple need not be regular, so that the bidual is not a Banach couple. The second difficulty is that if X is an interpolation space for $\overline{X}$, then, even if $\overline{X}$ is regular, the dual space X' need not be an intermediate space for $\overline{X}$, much less an interpolation space. A necessary condition for X' to be an intermediate space is that $\Delta\overline{X}$ be dense also in X. This does not, however, insure that X' is an interpolation space for $\overline{X}'$. The third difficulty is that there is no general rule for what a "dual method" should be for the construction of interpolation spaces, although general intuition and experience has usually led to the right constructions.

In this paper we are proposing a slightly different setting for interpolation theory. We propose to work in the somewhat larger category of doolittle diagrams (see Freyd [8] for the name) of Banach spaces, which we shall denote by $\overline{\mathfrak{B}}$ (mimicking the standard notation for the category of Banach couples). Our category $\overline{\mathfrak{B}}$ is the smallest (natural) category, containing the category of Banach couples while being closed under duality. $\overline{\mathfrak{B}}$ also enjoys some other interesting properties: it has a Ban-valued hom-functor (which simply means that the set of morphisms from $\overline{X}$ to $\overline{Y}$ is a Banach space under the natural norm) and also a very useful Ban-valued tensor product.

As we have just pointed out, the new setting of $\overline{\mathfrak{B}}$ takes care of the first of the traditional difficulties mentioned above. However, it would be naive to think that there are no difficulties inherent in this setting. The most important new difficulty that arises is that it is no longer completely obvious what an interpolation space should be. We have chosen to say that the "intersection" $\Delta\overline{X}$ (in our theory, the pullback) and the "sum-space" $\Sigma\overline{X}$ (the pushout) should be interpolation spaces, and then we consider two classes of interpolation spaces modelled on these paradigms. On the one hand we consider smaller (semi-) norms on $\Delta\overline{X}$ so that we get spaces that are completions of $\Delta\overline{X}$; we call such spaces Δ-interpolation spaces. On the other hand we consider larger (extended) norms on $\Sigma\overline{X}$ so that we get subspaces of $\Sigma\overline{X}$; these, which are the only interpolation spaces considered in the classical theory, we call Σ-interpolation spaces.

As a first result, we show that the most important classical methods, i.e. the real and the complex methods, have very natural definitions in our theory. Moreover, we show that even if some of them (the J-methods and to some extent the C_θ-method) are intrinsically Δ-methods, they actually turn out also to be Σ-interpolation methods.

When we begin to study duality questions in our theory, we run into part of the second traditional difficulty that unless $\Delta\overline{X}$ is dense in X, the dual space may be too large. Since this problem also exists in the $\overline{\mathfrak{B}}$-setting, our theory is quite satisfactory for Δ-interpolation spaces, while it is much less satisfactory for Σ-interpolation methods.

The other part of the second difficulty - that of insuring the duals of Δ-interpolation spaces for regular couples are interpolation spaces - is overcome in our theory with the aid of the tensor product. The problem arises because not all maps on a dual space are adjoints, so even if the space is preserved by all adjoints, it need not be preserved by all $\overline{X}'$-maps. Our tensor product makes it possible to consider a somewhat smaller substitute for the dual space which is preserved by $\overline{X}'$-maps (or in our terminology is a module over $L(\overline{X}') = L(\overline{X}',\overline{X}')$). Many of our results are formulated in terms of this "natural dual". The same definition can be applied also to Σ-interpolation spaces, but we have not been able to determine whether the "natural duals" are interpolation spaces in this case.

Finally, we come to the last difficulty, namely that there is no general rule for obtaining the "dual method" except that, as far as possible, it should give rise to the dual space when applied to the dual couple. We overcome this difficulty here by using the notion of a "dual functor", first defined by Fuks [9] and applied to Banach spaces by Mityagin and Švarc [19]. Since this notion is based on the "natural duality" between tensor products and hom-functors, which our category $\overline{\mathfrak{B}}$ is also endowed with, it is possible to define the dual functor for any Ban-valued functor F on $\overline{\mathfrak{B}}$. This dual functor has the property that the dual of a Δ-interpolation functor is a Σ-interpolation functor, while the dual of a Σ-functor is in some algebraic sense still some kind of interpolation functor. The most important classical methods that are not Δ-methods are the real $K(\theta,\infty)$-method and the complex C^{θ}-method.

The purpose of this paper is to construct a theory of interpolation which contains the classical theory and which is suitable for duality. We have not, however, tried to prove everything in the classical theory; e.g. we have made no efforts to prove compactness results or to generalize recent developments like the interpolation of more than two spaces or the notions of Calderón pairs or K-divisibility. On the other hand we have included the recent development by Janson [11] and Brudnyi-Krugljak [3] which has merged the important Aronszajn-Gagliardo paper [1] with the classical papers of Calderón [4] and Lions-Peetre [17] by showing that the real and complex methods are actually minimal methods in the sense of Aronszajn-Gagliardo. We have in fact strengthened these results somewhat by proving that the methods are not only minimal as interpolation functors but are minimal among all functors F such that for some couple $\overline{A}$, $F\overline{A} = A$.

In spite of the fact that one of the main features of our theory is the thesis that the ordinary Banach space dual is not the natural dual to consider in interpolation theory, we have made efforts to prove that in certain cases our dual is actually the ordinary dual. These efforts involve introducing a certain notion of "computability", for interpolation functors, which is related to the notion of computability for functors on Banach spaces introduced by Herz-Pelletier [10]. The "computable" interpolation functors behave much the way they are expected classically to behave.

Compared to most other papers on interpolation theory, even to those which are categorical "in spirit", ours is probably the most categorical. We have used several important ideas from category theory. To some extent this is unavoidable because the definition of the dual functor requires some sophisticated ideas from category theory. However, for the most part our use of category theory is intentional, because we feel that interpolation theory is so functorial in nature that category theory will lead to the correct notions. For example, we do feel that our notion of duality is the correct one for interpolation spaces while ordinary dual spaces are not sufficiently adapted to this situation. Along the same line of thinking, we feel that the notion of a Banach module (over the algebra $L(\overline{X})$) should be taken as the starting point for the idea of an interpolation space rather than the notion of an intermediate space. We believe that this more "algebraic" approach will be important for the study of questions arising out of interpolation theory, such as the "interpolation" of many (perhaps infinitely many) spaces.

Categorical methods have by now influenced most parts of Mathematics - except analysis. One reason why analysts have been unwilling to use categorical methods is probably that the languages of analysis and category theory are as difficult to translate as English and Swedish. In this paper we have occasionally had to choose either an analytical notation or a categorical one. We have made our compromise, and we hope that our paper will be readable for all.

Among our possibly diverse audience we anticipate that there will be functional analysts who merely wish to see that the real and complex methods are contained in a theory with better duality than hitherto present. We have organized the paper in such a way that Part I presents these results largely in terms of analysis. The experts in interpolation theory will, we hope, continue further in the paper to read about general interpolation functors in the $\overline{\mathfrak{B}}$-setting and their duality. We also expect that category theorists who are interested in applications of categorical methods to analysis will be among our readers. They may wish to begin directly with Part II, which is meant to be self-contained. In Chapter IV we have introduced and investigated all the categorical properties that are natural in $\overline{\mathfrak{B}}$, not all of which are actually used in the paper. We have also tried to indicate from time to time the extent of categorical generality inherent in our constructions in the hope that applications of these ideas may arise in other areas. Finally, we hope our readership will include mathematicians who are interested in the interplay of various branches of Mathematics. They should be particularly interested in Part III, which is our attempt to tie together the concrete applications of Chapters II and III with the more abstract theory of Chapters VI and VII.

Parts of this work have been presented previously. The real method à la Chapter II was presented at a conference on interpolation theory held in Lund, Sweden in August 1983, and a paper [14] based on this presentation is contained in the Conference Proceedings. Preliminary versions of the more categorical aspects of the paper have been presented at conferences in Sussex, England, Denver, Colorado and Murten, Switzerland, and the articles [15] and [23] have emerged.

In closing we wish to make some acknowledgments. Several institutions have hosted us during some part of our four years' collaboration and we are grateful for their hospitality: our home universities - York and Uppsala - the University of Connecticut, and McGill University. We wish to single out for thanks Ms. P. Ferguson of McGill University and Ms. Paula Panaro of York University for their superb typing of the preliminary and final versions of the manuscript, respectively. We are grateful to the Natural Sciences and Engineering Research Council for its support, without which this collaboration would not have been possible. We also acknowledge the interest and encouragement of several mathematicians, in particular, J.W. Gray, J. Peetre, and S. Janson. Finally, we thank C. Herz, who, anticipating our common interests, introduced us to one another.

PART I

CHAPTER I

PRELIMINARIES

1. The Setting.

As we have explained at length in the introduction, we feel that the category of Banach couples is not the best setting for interpolation theory. We are proposing to work in a larger category - the category of doolittle diagrams of Banach spaces - which is a simple extension of the category of Banach couples enjoying the property of being closed under duality. We believe that, despite certain difficulties arising in this setting, this is the "right category" for studying interpolation.

We begin by giving our basic definitions.

1.1 Definition. A doolittle diagram $\overline{X}$ of Banach spaces is a commutative diagram (of Banach spaces)

$$\overline{X} = \begin{array}{ccc} \Delta\overline{X} & \xrightarrow{\delta_0} & X_0 \\ {\scriptstyle\delta_1}\downarrow & & \downarrow{\scriptstyle\sigma_0} \\ X_1 & \xrightarrow[\sigma_1]{} & \Sigma\overline{X} \end{array}$$

such that

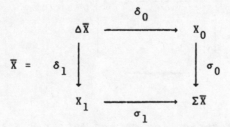

(i) all maps are continuous linear maps and

(ii) $\overline{X}$ is both a pullback and pushout.

Condition (ii) means the above diagram is commutative and that $\Delta\overline{X}$ and $\Sigma\overline{X}$ are "universal" in the following sense: if Y is a "candidate" for the top left corner, i.e. if there are maps $f_i: Y \to X_i$, $i=0,1$, such that $\sigma_0 \circ f_0 = \sigma_1 \circ f_1$, then Y factors uniquely through $\Delta\overline{X}$, i.e. there is a unique map $f: Y \to \Delta\overline{X}$ such that $f \circ \delta_i = f_i$, $i=0,1$, and similarly for $\Sigma\overline{X}$.

In practice we can give the following concrete description of doolittle diagrams of Banach spaces. First for a pair (X_0, X_1) of Banach spaces we denote by $X_0 \pi X_1$ and $X_0 \amalg X_1$ the product and sum (or coproduct) spaces, respectively, where

$$\|(x_0, x_1)\|_{X_0 \pi X_1} = \sup(\|x_0\|, \|x_1\|)$$

and

$$\|x\|_{X_0 \amalg X_1} = \inf(\|x_0\| + \|x_1\| \,|\, x = x_0 + x_1) .$$

1.2. <u>Proposition</u>. A doolittle diagram of Banach spaces is determined by a pair (X_0, X_1) of Banach spaces and a closed subspace $\Delta\overline{X}$ of $X_0 \pi X_1$.

<u>Proof</u>: Let

$$D = \begin{array}{ccc} P & \xrightarrow{\;u\;} & X_0 \\ {\scriptstyle v}\downarrow & & \downarrow{\scriptstyle f} \\ X_1 & \xrightarrow[\;g\;]{} & Q \end{array}$$

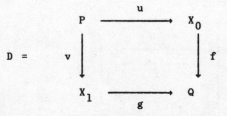

be a doolittle diagram of Banach spaces and let $\varphi: P \to X_0 \pi X_1$ be defined by $\varphi = (u, v)$. Then by the definition of the pullback, one sees that $\wp$ is an isometry, so P may be considered a closed subspace of $X_0 \pi X_1$. More precisely, we see that P is isomorphic to the subset of $X_0 \pi X_1$ consisting of those (x_0, x_1) such that $fx_0 = gx_1$.

Conversely, if $\Delta\overline{X}$ is a closed subspace of $X_0 \pi X_1$, we denote by $\delta_i : \Delta\overline{X} \to X_i$, the projection of $\Delta\overline{X}$ to X_i. Then the pushout Q in the diagram

$$\begin{array}{ccc} \Delta\overline{X} & \xrightarrow{\ \delta_0\ } & X_0 \\ {\scriptstyle \delta_1}\downarrow & & \downarrow{\scriptstyle \sigma_0} \\ X_1 & \xrightarrow[\ \sigma_1\ ]{} & Q \end{array}$$

can be described as a quotient of $X_0 \amalg X_1$ over the subspace

$$\Delta\overline{X}^{-} = \{(x_0, x_1) \mid \exists x \in \Delta\overline{X},\ x_0 = \delta_0 x,\ x_1 = -\delta_1 x\};$$

$\sigma_i : X_i \to Q$ are the canonical maps. It is easy to verify that $\Delta\overline{X}$ is the pullback of the above diagram, and hence, that it is a doolittle diagram. □

The general doolittle diagram in our paper will be denoted by $\overline{X}$ or $(X_0, X_1, \Delta\overline{X})$, where $\Delta\overline{X}$ is understood to be a closed subspace of $X_0 \pi X_1$, and it will be equipped with morphisms as follows:

$$\begin{array}{ccc} \Delta\overline{X} & \xrightarrow{\ \delta_0\ } & X_0 \\ {\scriptstyle \delta_1}\downarrow & & \downarrow{\scriptstyle \sigma_0} \\ X_1 & \xrightarrow[\ \sigma_1\ ]{} & \Sigma\overline{X} \end{array} .$$

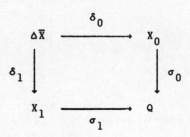

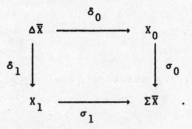

Since the diagram is commutative, we have $\sigma_0 \circ \delta_0 = \sigma_1 \circ \delta_1$. We shall denote this frequently used map by j and call $\overline{X}$ <u>non-trivial</u> if $j \neq 0$.

1.3. <u>Examples</u>. 1. A sum diagram and a product diagram

$$\begin{array}{ccc} 0 & \longrightarrow & X_0 \\ \big\downarrow & & \big\downarrow \\ X_1 & \longrightarrow & X_0 \amalg X_1, \end{array} \qquad\qquad \begin{array}{ccc} X_0 \Pi X_1 & \longrightarrow & X_0 \\ \big\downarrow & & \big\downarrow \\ X_1 & \longrightarrow & 0 \end{array}$$

are both (trivial) doolittle diagrams. 2. Every Banach couple is a doolittle diagram such that all the maps are injective; conversely, a doolittle diagram the maps of which are injective is simply a Banach couple.

Since the purpose of interpolation theory is to interpolate operators, we have to know what an operator between doolittle diagrams is.

1.4. <u>Definition</u>. Let $\overline{X}$ and $\overline{Y}$ be doolittle diagrams. A map T from $\overline{X}$ to $\overline{Y}$ is a pair (T_0, T_1) of continuous linear maps such that the following diagram commutes:

$$\begin{array}{ccccccc} & & X_0 & \xrightarrow{T_0} & Y_0 & & \\ & \overset{\delta_0}{\nearrow} & & & & \overset{\sigma_0}{\searrow} & \\ \Delta\overline{X} & & & & & & \Sigma\overline{Y} \\ & \underset{\delta_1}{\searrow} & & & & \underset{\sigma_1}{\nearrow} & \\ & & X_1 & \xrightarrow[T_1]{} & Y_1 & & \end{array}$$

(We are deliberately avoiding notation like $\delta_0(\overline{X})$, $\sigma_1(\overline{Y})$ which is cumbersome.)

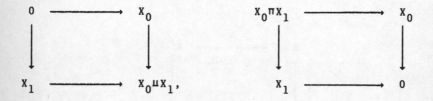

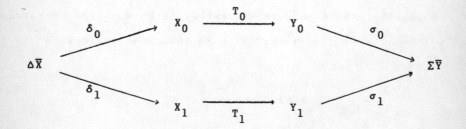

1.5. <u>Remarks</u>. 1. We note that when $\overline{X}$ and $\overline{Y}$ are Banach couples, our definition of morphism is the same as the classical definition.

2. In view of the definition of the pullback, the map $\sigma_0 \circ T_0 \circ \delta_0 = \sigma_1 \circ T_1 \circ \delta_1$ factors through $\Delta\overline{Y}$, so there exists $\Delta T: \Delta\overline{X} \to \Delta\overline{Y}$. Similarly, from the definition of the pushout we get $\Sigma T: \Sigma\overline{X} \to \Sigma\overline{Y}$.

We shall denote by $L(\overline{X},\overline{Y})$ the set of all maps from $\overline{X}$ to $\overline{Y}$. Actually, $L(\overline{X},\overline{Y})$ is a Banach space under the norm

$$\|T\| = \max(\|T_0\|, \|T_1\|).$$

We may also observe from our description of pullbacks given in 1.2 that $L(\overline{X},\overline{Y})$ is the pullback of the diagram

$$\begin{array}{ccc} & & L(X_0,Y_0) \\ & & \downarrow \\ L(X_1,Y_1) & \longrightarrow & L(\Delta\overline{X},\Sigma\overline{Y}). \end{array}$$

The category of doolittle diagrams of Banach spaces and bounded linear morphisms as described above is denoted by $\overline{\mathfrak{B}}$, while the subcategory of Banach couples is denoted by $\overline{\mathfrak{B}\mathfrak{C}}$; $\mathfrak{B}$ will denote the category of Banach spaces.

Since we have motivated our introduction of $\overline{\mathfrak{B}}$ by a discussion of the better duality properties it enjoys, we should begin at least by showing that $\overline{\mathfrak{B}}$ is closed under duals.

1.6 <u>Proposition</u>. Let $\overline{X}$ be a doolittle diagram and let $\overline{X}'$ be the diagram

$$\begin{array}{ccc}
(\Sigma\overline{X})' & \xrightarrow{\sigma_0'} & X_0' \\
\downarrow{\scriptstyle \sigma_0'} & & \downarrow{\scriptstyle \delta_0'} \\
X_1' & \xrightarrow[\delta_1']{} & (\Delta\overline{X})' .
\end{array}$$

Then $\overline{X}'$ is a doolittle diagram, i.e. $\Delta\overline{X}'=(\Sigma\overline{X})'$ and $\Sigma\overline{X}'=(\Delta\overline{X})'$.

<u>Proof</u>: The commutativity of the above diagram is obvious. That it is a pullback follows from the pushout property of $\overline{X}$ directly. That it is a pushout as well is a fact, non-trivial only in the sense that it depends on a deep theorem, namely the Hahn-Banach theorem. □

2. <u>Doolittle diagrams, Couples, and Regular Couples</u>.

We have observed above that a Banach couple is merely a doolittle diagram such that all maps are injective. The main difference then between our category $\overline{\mathfrak{B}}$ and the traditional category $\overline{\mathfrak{BC}}$ is that in an arbitrary doolittle diagram the maps need not be injective. It is natural, therefore, to consider the kernels of the maps δ_i, σ_i in $\overline{X}$, at least one of which will be a non-trivial space if $\overline{X}$ is not a Banach couple.

Let us denote by $K_i\overline{X}$ the space $\ker(\sigma_i)$ $(\subset X_i)$. We may prove the following proposition.

2.1. <u>Proposition</u>. Let $\overline{X}$ be a doolittle diagram. Then $\ker(\delta_0) = \ker(\sigma_1)(=K_1\overline{X})$ and $\ker(\delta_1) = \ker(\sigma_0)(=K_0\overline{X})$.

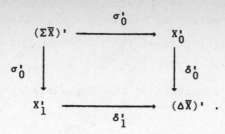

<u>Proof</u>: Recall from 1.2 that $\Delta\overline{X}$ may be interpreted as the following subspace of $X_0 \sqcap X_1$:

$$\{(x_0,x_1) \mid \sigma_0 x_0 = \sigma_1 x_1\}.$$

Let $x_1 \in K_1\overline{X}$. Then $\sigma_1 x_1 = 0$, so $x = (0,x_1) \in \Delta\overline{X}$. Therefore, $\delta_0 x = 0$, so $K_1\overline{X} \subset \ker(\delta_0) \subset \Delta\overline{X}$. Conversely, if $y \in \ker(\delta_0)$, then $y = (0,y_1)$, where y_1 is such that $\sigma_1 y_1 = 0$, so $y_1 \in K_1\overline{X}$. Hence, $\ker(\delta_0) \subset K_1\overline{X}$. The same argument proves that $K_0\overline{X} = \ker(\delta_1)$. □

We shall write $K\overline{X} = K_0\overline{X} \sqcap K_1\overline{X} \subset \Delta\overline{X}$ and observe that $K\overline{X} = \ker(j)$. $\overline{K\overline{X}}$ will denote the trivial doolittle diagram

$$\overline{K\overline{X}} = \begin{array}{ccc} K\overline{X} & \longrightarrow & K_0\overline{X} \\ \downarrow & & \downarrow \\ K_1\overline{X} & \longrightarrow & 0 \end{array} .$$

Letting $Y_i = X_i/K_i\overline{X}$, we can define a pullback diagram

$$\begin{array}{ccc} \Delta\overline{Y} & \longrightarrow & Y_0 \\ \downarrow & & \downarrow \\ Y_1 & \longrightarrow & \Sigma\overline{X} \end{array} .$$

Then by the property of the pullback, $j: \Delta\overline{X} \to \Sigma\overline{X}$ must factor through $\Delta\overline{Y} \to \Sigma\overline{X}$, which is obviously injective. It is easy to conclude that $\Delta\overline{Y} = \Delta\overline{X}/K\overline{X}$ and that

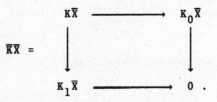

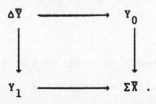

$$\bar{X}/\bar{K}\bar{X} = \begin{array}{ccc} \Delta\bar{X}/K\bar{X} & \longrightarrow & X_0/K_0\bar{X} \\ \downarrow & & \downarrow \\ X_1/K_1\bar{X} & \longrightarrow & \Sigma\bar{X} \end{array}$$

is a doolittle diagram, in fact a Banach couple.

Now, let $T = (T_0, T_1): \bar{X} \to \bar{Y}$ be a map of doolittle diagrams. If $x_0 \in K_0\bar{X}$, then $x = (x_0, 0) \in \Delta\bar{X}$, so $(\sigma_1 \circ T_1 \circ \delta_1)(x) = 0 = (\sigma_0 \circ T_0 \circ \delta_0)(x)$, and, hence, $T_0 x_0 \in K_0\bar{Y}$. Similarly, $T_1 x_1 \in K_1\bar{Y}$ if $x_1 \in K_1\bar{X}$. In categorical terms this means that "K_i" and "$()/_{\bar{K}()}$" are functors from $\bar{\mathfrak{B}}$ to $\mathfrak{B}$ and $\bar{\mathfrak{B}}$ to $\bar{\mathfrak{B}}\mathfrak{C}$, respectively. It also follows that if $\bar{X}$ is a doolittle diagram and $\bar{Y}$ a Banach couple, then every morphism $T: \bar{X} \to \bar{Y}$ factors through $\bar{X}/\bar{K}\bar{X}$.

We recall that a sequence

$$\to X \xrightarrow{u} Y \xrightarrow{v} Z \to$$

of Banach spaces is said to be exact at Y if $\mathrm{im}(u) = \ker(v)$. In analogy we shall say that a sequence

$$\to \bar{X} \to \bar{Y} \to \bar{Z} \to$$

is exact at $\bar{Y}$ if it is exact at each vertex of $\bar{Y}$. In particular it follows that for every doolittle diagram $\bar{X}$ we have a short exact sequence

$$\bar{0} \to \bar{K}\bar{X} \to \bar{X} \to \bar{X}/\bar{K}\bar{X} \to \bar{0} ,$$

where $\bar{0}$ denotes the diagram with the 0-space at each vertex.

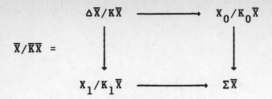

In the classical duality theorems for the real and the complex methods it is assumed that a given Banach couple is a regular Banach couple in the sense that $\Delta\overline{X}$ is dense in X_0 and X_1. If $\overline{X}$ is not regular, then it is sometimes possible to replace $\overline{X} = (X_0, X_1)$ by $\overline{X}^0 = (X_0^0, X_1^0)$, where X_i^0 is the closure of the image of δ_i in X_i. If $\overline{X}$ is any doolittle diagram, the same construction will be used to form

$$\overline{X}^0 = \begin{array}{ccc} \Delta\overline{X} & \longrightarrow & X_0^0 \\ \downarrow & & \downarrow \\ X_1^0 & \longrightarrow & \Sigma\overline{X}^0 \end{array},$$

where $\Sigma\overline{X}^0$ is the pushout. Then for every doolittle diagram there exists a short exact sequence

$$\overline{0} \to \overline{X}^0 \to \overline{X} \to \overline{X}/\overline{X}^0 \to \overline{0},$$

where $\overline{X}/\overline{X}^0$ is the sum diagram

$$\begin{array}{ccc} 0 & \longrightarrow & X_0/X_0^0 \\ \downarrow & & \downarrow \\ X_1/X_1^0 & \longrightarrow & \left[X_0/X_0^0\right] \amalg \left[X_1/X_1^0\right]. \end{array}$$

Finally, we note without proof the straightforward proposition concerning the operations $\overline{X} \mapsto \overline{X}^0$ and $\overline{X} \mapsto \overline{X}/\overline{K}\overline{X}$, which is useful when moving from doolittle diagrams to regular Banach couples.

2.2. <u>Proposition</u>. The operations $\overline{X} \mapsto \overline{X}^0$ and $\overline{X} \mapsto \overline{X}/\overline{K}\overline{X}$ are commutative, i.e.

$$(\overline{X}/\overline{K}\overline{X})^0 = \overline{X}^0/(\overline{K}\overline{X}).$$

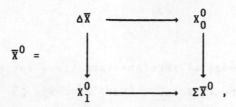

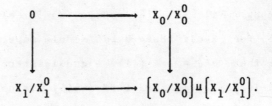

Proof: This follows easily from the fact that $K_i\overline{X} \subset \mathrm{im}(\delta_i)$, i=0,1. □

3. Interpolation Spaces.

In the classical theory an interpolation space for a Banach couple $\overline{X}$ is a Banach space X such that $\Delta\overline{X} \subset X \subset \Sigma\overline{X}$ ("$\subset$" simply means a monomorphic inclusion) and such that each map $T:\overline{X} \to \overline{X}$ restricts to a bounded linear map $T:X \to X$. (We shall always speak of the second condition by saying that X is required to be an $L(\overline{X})$-module, where $L(\overline{X})$ is our abbreviation for $L(\overline{X},\overline{X})$.) In the $\overline{\mathfrak{B}}$-setting we do not have an inclusion $\Delta\overline{X} \subset \Sigma\overline{X}$, so we begin by requiring only the condition that X be an $L(\overline{X})$-module. We proceed as follows:

3.1. Definitions. (i) An $L(\overline{X})$-module X is called a quasi-interpolation space for $\overline{X}$ if there exist module maps $\delta:\Delta\overline{X} \to X$ and $\sigma:X \to \Sigma\overline{X}$ such that $\sigma\circ\delta = j$; (ii) A quasi-interpolation space X is called a Δ-interpolation space if $\mathrm{im}(\delta)$ is dense in X; (iii) A quasi-interpolation space X is called a Σ-interpolation space if σ is a monomorphism.

Although we shall leave the precise definition of interpolation functors to Chapter IV, let us say here that an interpolation functor is just a "method" of obtaining (some kind of) interpolation space from a doolittle diagram $\overline{X}$. The real and complex methods of interpolation are, of course, functors.

We have indicated in our introduction some reasons for the usefulness of our two types of interpolation spaces. However, most examples known to us of Δ-interpolation spaces (besides Δ itself) are actually Σ-interpolation spaces as well. After wondering if this is always the case, we came up with the following example of a Δ-interpolation space (unequal to $\Delta\overline{X}$) which is not a Σ-interpolation space.

3.2. <u>Example</u>. Let $\ell^p(\underline{n})$ denote the n-dimensional real ℓ^p-space. Let $\overline{X}$ be the doolittle diagram

$$\begin{array}{ccc} \ell^1(\underline{2})\Pi\mathbb{R} & \xrightarrow{\ \delta_0\ } & \ell^\infty(\underline{3}) \\ {\scriptstyle\delta_1}\downarrow & & \downarrow{\scriptstyle\sigma_0} \\ \ell^1(\underline{2}) & \xrightarrow[\ \sigma_1\ ]{} & \ell^\infty(\underline{2}) \end{array} ,$$

where δ_1 and σ_0 are the projections on the first two coordinates and where δ_0 and σ_1 are the canonical embeddings. Then $\ell^2(\underline{2})\Pi\mathbb{R}$ (which is obtained, for example, by applying the real $(1/2,2)$-method to the couple $(\ell^\infty(\underline{3}),\ell^1(\underline{2})\Pi\mathbb{R}))$ is a Δ-interpolation space for $\overline{X}$ which is not a Σ-interpolation space.

CHAPTER II

THE REAL METHOD

1. The J- and K- methods.

Abstract interpolation theory was developed around 1960 as a generalization of the famous theorems of Riesz-Thorin and Marcinkiewicz (see [2]). The abstract version of the former theorem gave rise to a now standard construction which is called the "complex method of interpolation"; this will be discussed in the next chapter. The Marcinkiewicz theorem, on the other hand, inspired several different generalizations, the various versions being presented around 1960. The introduction of the J- and K- functionals by Lions-Peetre [17] and their subsequent "equivalence theorem", which implies that several possible constructions are in fact all equivalent, brought some clarity to the situation, and justified applying the term "real method" to these equivalent constructions.

We shall start by defining the J- and K- functionals and methods in the setting of the category $\overline{\mathfrak{B}}$ introduced in Chapter I.

1.1. Definition. Let $\overline{X}$ be a doolittle diagram. Then the functionals J and K are defined on $\Delta\overline{X}$ and $\Sigma\overline{X}$, respectively, by

$$J(t,x) = \max(\|\delta_0 x\|, t\|\delta_1 x\|)$$

and

$$K(t,x) = \inf(\|x_0\| + t\|x_1\| \mid \sigma_0 x_0 + \sigma_1 x_1 = x),$$

for a given positive real number t.

The J- and K- functionals satisfy the usual standard properties as given by the following proposition. (We recommend the text by Bergh and Löfström [2] as a reference for the classical exposition of this material.)

1.2. Proposition. $J(t,x)$ is a positive, increasing, convex function of t, while $K(t,x)$ is positive, increasing, and concave. Furthermore, we have the following well known inequalities:

(i) $J(t,x) \leq \max(1,t/s)J(s,x)$, $x \in \Delta\overline{X}$,

(ii) $K(t,x) \leq \max(1,t/s)K(s,x)$, $x \in \Sigma\overline{X}$,

(iii) $K(t,jx) \leq \min(1,t/s)J(s,x)$, $x \in \Delta\overline{X}$.

Moreover, J and K satisfy the following duality properties, which simply state that the J- and K- functionals are dual to each other.

1.3. Proposition. Let $\overline{X}$ be a doolittle diagram with dual diagram $\overline{X}'$ as defined in I.1.6. If $x \in \Delta\overline{X}$ and $x' \in \Sigma\overline{X}'$, then we have

(i) $\langle x',x\rangle \leq K(t,x')J(1/t,x)$,

and in fact

(ii) $K(t,x') = \sup\{|\langle x',x\rangle| \,\big|\, J(1/t,x) \leq 1\}$ and

(iii) $J(t,x) = \sup\{|\langle x',x\rangle| \,\big|\, K(1/t,x') \leq 1\}$.

Finally, the corresponding statements are true with $x \in \Sigma\overline{X}$ and $x' \in \Delta\overline{X}'$ as well.

We are now in a position to define the J- and K- methods of interpolation. We shall begin by defining the J-method. We note that although our construction is conceptually different from the classical method, it will nevertheless prove to be equivalent.

First, we let $C_c(\mathbb{R}^+,\Delta\bar{X})$ denote the space of continuous $\Delta\bar{X}$-valued functions with compact support on the interval $(0,\infty)$. There is a homogeneous sublinear map

$$m: C_c(\mathbb{R}^+,\Delta\bar{X}) \to C_c(\mathbb{R}^+,\mathbb{R})$$

defined by

$$m(u(t)) = J(t,u(t)),$$

for $u \in C_c(\mathbb{R}^+,\Delta\bar{X})$ and $0 < t < \infty$.

On the space $C_c(\mathbb{R}^+,\mathbb{R})$ we have a two-parameter family of norms $\Phi_{\theta,q}$ given by

$$\Phi_{\theta,q}(f) = \|t^{-\theta}f\|_{L^q(dt/t)}, \text{ for } 0 < \theta < 1,\ 1 \le q \le \infty,$$

i.e.

$$\Phi_{\theta,q}(f) = \left[\int_0^\infty |t^{-\theta}f(t)|^q \frac{dt}{t}\right]^{1/q} \quad \text{for } 1 \le q < \infty$$

and

$$\Phi_{\theta,\infty}(f) = \sup|t^{-\theta}f(t)|.$$

We also have a natural $\Delta\bar{X}$-valued integral on $C_c(\mathbb{R}^+,\Delta\bar{X})$ given by

$$\int_0^\infty u(t)\,\frac{dt}{t}, \text{ for } u \in C_c(\mathbb{R}^+,\Delta\bar{X}).$$

In terms of the preceding notations it is clear that the functional $J(\theta,q,-)$ defined on $\Delta\bar{X}$ by

$$(*) \qquad J(\theta,q,x) = \inf\{\Phi_{\theta,q}(m(u)) \mid \int_0^\infty u(t)\,\frac{dt}{t} = x,\ u \in C_c(\mathbb{R}^+,\Delta\overline{X})\}$$

is a semi-norm.

1.4. <u>Definition</u>. Let $\overline{X}$ be a doolittle diagram. Then the J-interpolation space $J(\theta,q,\overline{X})$ is defined as the completion of the space $\Delta\overline{X}$ with respect to the seminorm $J(\theta,q,-)$ defined by (*) above.

1.5. <u>Remark</u>. The discrete $J(\theta,q)$-method is defined as the completion of $\Delta\overline{X}$ with respect to the corresponding discrete $(\theta,q,)$-norm:

$$\|x\| = \inf\{(\sum_{-\infty}^{\infty}(2^{-k\theta}J(2^k,u_k))^q)^{1/q} \mid x = \sum_{-\infty}^{\infty} u_k,\ u_k \in \Delta\overline{X}\}.$$

The following proposition is clear from our construction and the usual classical calculations.

1.6. <u>Proposition</u>. The function which to every doolittle diagram $\overline{X}$ associates the space $J(\theta,q,\overline{X})$ is a functor from $\overline{\mathfrak{B}}$ to $\mathfrak{B}$. Moreover, for every $T = (T_0,T_1) \in L(\overline{X},\overline{Y})$ the map $\Delta T:\Delta\overline{X} \to \Delta\overline{Y}$ extends by continuity to a continuous map $T:J(\theta,q,\overline{X}) \to J(\theta,q,\overline{Y})$ such that $\|T\| \le \|T_0\|^{1-\theta}\|T_1\|^{\theta}$.

While our presentation of the J-method differs from the classical presentation, the K-method is defined exactly as in the classical case.

1.7. <u>Definition</u>. Let $\overline{X}$ be a doolittle diagram. The K-interpolation space $K(\theta,q,\overline{X})$ is defined as the set

$$\{x \in \Sigma\overline{X} \mid \Phi_{\theta,q}(K(t,x)) < \infty\}$$

with the obvious norm for $0 < \theta < 1$, $1 \le q \le \infty$.

We leave it to the reader to verify that the space $K(\theta,q,\overline{X})$ is, like $J(\theta,q,\overline{X})$, an "exact interpolation space of exponent θ".

1.8. Remarks. (1) The discrete $K(\theta,q)$-method is defined correspondingly. (2) It should be pointed out that the discrete methods are equivalent to the "continuous methods" up to a factor of 2^{θ}, so that we shall feel free to replace the continuous methods by the discrete whenever it is necessary (or just convenient).

2. The Duality Theorem.

The duality theorem for the J- and K- methods is an important theorem of interpolation theory. However, in the classical theory the need to restrict this result to the category of regular Banach couples has made it less useful than it could have been. In our setting, however, this restriction is not necessary. In fact our new definition of the J-method nearly trivializes one half of the duality theorem which follows.

2.1. Theorem. Let $\overline{X}$ be a doolittle diagram. Let $0 < \theta < 1$ and let q' be such that $1/q + 1/q' = 1$. Then

$$J(\theta,q,\overline{X})' = K(\theta,q',\overline{X}'), \quad \text{if } 1 \le q \le \infty,$$
$$K(\theta,q,\overline{X})' = J(\theta,q',\overline{X}'), \quad \text{if } 1 \le q < \infty,$$

and

$$K^0(\theta,\infty,\overline{X})' = J(\theta,1,\overline{X}'),$$

where

$$K^0(\theta,\infty,\overline{X}) = \{x \in K(\theta,\infty,\overline{X}) \,|\, t^{-\theta}K(t,x) \to 0 \quad \text{as} \quad t \to 0 \quad \text{or} \quad t \to \infty\}.$$

In this section we shall prove the first half of the above theorem, which we restate as Proposition 2.2. The remainder of the theorem will be a consequence of the results of Section 3.

2.2. Proposition. If $\overline{X}$ is a doolittle diagram, then

$$J(\theta,q,\overline{X})' = K(\theta,q',\overline{X}'),$$

for $0 < \theta < 1$ and $1 \leq q \leq \infty$.

Proof: Since $J(\theta,q,\overline{X})$ is the completion of $\Delta\overline{X}$ with respect to a smaller (semi-)norm, the dual space consists of those elements of $(\Delta\overline{X})'=\Sigma\overline{X}'$ that happen to be continuous with respect to this smaller norm. Therefore, let $x'\in\Sigma\overline{X}'$, and suppose that $x' \in K(\theta,q',\overline{X}')$ with $\|x'\|_{K(\theta,q',\overline{X}')}= 1$, i.e.

$$\left[\int_0^\infty |t^{-\theta}K(t,x')|^{q'} \frac{dt}{t}\right]^{1/q'} = 1,$$

for $1 \leq q' < \infty$, and

$$\sup|t^{-\theta}K(t,x')| = 1$$

if $q' = \infty$. Now if $u \in C_c(\mathbb{R}^+, \Delta\overline{X})$ and if $x = \int_0^\infty u(t)\,\frac{dt}{t}$, then for $1 < q < \infty$, we have by using Proposition 1.3,

$$\begin{aligned}\langle x', x\rangle &= \int_0^\infty \langle x', u(t)\rangle dt/t \le \int_0^\infty K(1/t, x')J(t, u(t))dt/t \\ &\le \int_0^\infty (t^{\theta}K(1/t, x'))(t^{-\theta}J(t, u(t)))dt/t \\ &\le \left[\int_0^\infty |t^{\theta}K(1/t, x')|^{q'}\,\frac{dt}{t}\right]^{1/q'} \left[\int_0^\infty |t^{-\theta}J(t, u(t))|^{q}\,\frac{dt}{t}\right]^{1/q} \\ &= \left[\int_0^\infty |t^{-\theta}K(t, x')|^{q'}\,\frac{dt}{t}\right]^{1/q'} \Phi_{\theta,q}(m(u)) \\ &= \|x'\|_{K(\theta,q',\overline{X}')}\Phi_{\theta,q}(m(u)).\end{aligned}$$

Therefore, since this holds for every representation of x, it follows that for all $x \in \Delta\overline{X}$,

$$|\langle x', x\rangle| < \|x'\|_{K(\theta,q',\overline{X}')}\|x\|_{J(\theta,q,\overline{X})}.$$

This proves that x' belongs to $J(\theta,q,\overline{X})'$ and as such has norm ≤ 1.

To prove that $\|x'\|_{J(\theta,q,\overline{X})'} = \|x'\|_{K(\theta,q',\overline{X}')}$ we have to use Proposition 1.3 to construct a suitable function in $C_c(\mathbb{R}^+, \Delta\overline{X})$. In fact it is possible to choose a function $u(t) \in C_c(\mathbb{R}^+, \Delta\overline{X})$ such that

$$\int_0^\infty |t^{-\theta}J(t, u(t))|^q\,\frac{dt}{t} \le 1$$

and such that for any $a < 1$ and $N > 1$, we have

$$\int_0^\infty \langle x', u(t)\rangle\,\frac{dt}{t} \ge a\left[\int_{1/N}^N |t^{-\theta}K(t, x')|^{q'}\,\frac{dt}{t}\right]^{1/q'},$$

and this proves that $\|x'\|_{J(\theta,q,\overline{X})'} = \|x'\|_{K(\theta,q',\overline{X}')}$.

We observe that with suitable modifications the above reasoning also holds in the extreme cases $q=1$ or ∞. □

2.3. Remark. The above theorem shows that the J- and K- methods are dual in the usual concrete sense. In Part II of our work we define the notion of a "dual functor" associated to each functor from $\overline{\mathfrak{B}}$ to $\mathfrak{B}$, and we show that the duals of interpolation functors (in a sense to be made precise) are interpolation functors. It evolves, moreover, that the J- and K- methods of interpolation are the dual functors of one another.

3. The Equivalence Theorem.

The first step in the comparison of the J- and K- methods is given by the following proposition.

3.1. Proposition. The canonical map $j = \sigma\circ\delta:\Delta\overline{X} \to \Sigma\overline{X}$ factors in a natural way as $\Delta\overline{X} \xrightarrow{\delta} J(\theta,q,\overline{X}) \xrightarrow{\varphi} K(\theta,q,\overline{X}) \xrightarrow{\sigma} \Sigma\overline{X}$, where φ is continuous with $\|\varphi\| \le [\theta(1-\theta)]^{-1}$.

Proof: We must first prove that $\operatorname{im}(j) \subset K(\theta,q,\overline{X})$. Let $x\in\Delta\overline{X}$ with $\|x\|_{\Delta\overline{X}} = 1$. Then since $\|x\|_{\Delta\overline{X}} = J(1,x)$, we have by Proposition 1.2 that

$$K(t,jx) \le \min(1,t).$$

Therefore,

$$\|jx\|_{K(\theta,q,\overline{X})} \le \|t^{-\theta}\min(1,t)\|_{L^q(\frac{dt}{t})} = \left[\frac{1}{q\theta(1-\theta)}\right]^{1/q} = C_{\theta,q},$$

for $1 \le q < \infty$, and

$$\|jx\|_{K(\theta,\infty,\overline{X})} \le 1 = C_{\theta,\infty},$$

so $jx \in K(\theta,q,\overline{X})$.

Next we prove that $j:\Delta\overline{X} \to K(\theta,q,\overline{X})$ is continuous when $\Delta\overline{X}$ is normed by the seminorm $J(\theta,q)$. To this end let $u(t) \in C_c(\mathbb{R}^+,\Delta\overline{X})$ and let $x = \int_0^\infty u(t)dt/t$. Then we have

$$K(t,jx) = K(t,\int_0^\infty u(s)ds/s) \le \int_0^\infty K(t,j(u(s)))\,\frac{ds}{s}.$$

Using Proposition 1.2 again, we have that

$$K(t,j(u(s))) \le \min(1,t/s)J(s,u(s)).$$

Therefore, if $1 \le q < \infty$, we have

$$\left[\int_0^\infty |t^{-\theta}K(t,jx)|^q\,\frac{dt}{t}\right]^{1/q} \le \left[\int_0^\infty \left[t^{-\theta}\int_0^\infty K(t,j(u(s)))\,\frac{ds}{s}\right]^q\frac{dt}{t}\right]^{1/q}$$
$$\le \left[\int_0^\infty \left[t^{-\theta}\int_0^\infty \min(1,t/s)J(s,u(s))\,\frac{ds}{s}\right]^q\frac{dt}{t}\right]^{1/q}.$$

Using the change of variables $s'=t/s$ and Minkowski's inequality, we may continue the above inequalities as follows:

$$= \int_0^\infty\left[t^{-\theta}\int_0^\infty \min(1,s')J(t/s',u(t/s'))\,\frac{ds'}{s'}\right]^q\frac{dt}{t}\Bigg]^{1/q}$$
$$\le \int_0^\infty \min(1,s')(s')^{-\theta}\left[\int_0^\infty |(t/s')^{-\theta}J(t/s',u(t/s'))|^q\,\frac{dt}{t}\right]^{1/q}\frac{ds'}{s'}$$
$$= \frac{1}{\theta(1-\theta)}\,\Phi_{\theta,q}(m(u)).$$

Since this inequality holds for every representation $x = \int_0^\infty u(t)dt/t$, it follows that

$$\|jx\|_{K(\theta,q,\overline{X})} \le \frac{1}{\theta(1-\theta)} \|x\|_{J(\theta,q,\overline{X})}.$$

If $q=\infty$, then the same estimate holds and is easier. Therefore, j extends to $J(\theta,q,\overline{X})$ by continuity and this gives the map φ required in the proposition. □

The next step in verifying the equivalence of the J- and K-methods is showing that the map φ is surjective. For this we shall, as in the classical case, use the discrete J- and K- methods.

We formulate our version of the "fundamental lemma of interpolation" as follows.

3.2. Proposition. Let $x \in \Sigma\overline{X}$. Then there exists a sequence $\{u_k\}_{k\in Z}$, $u_k \in \Delta\overline{X}$, such that

(i) $J(2^k,u_k) \le 4K(2^k,x)$ and

(ii) $\|x-j(\sum_{n+1}^{m} u_k)\|_{\Sigma\overline{X}} \le \min(1,2^{-m})K(2^m,x) + \min(1,2^{-n})K(2^n,x)$,

for $n < 0 < m$.

Before proving this proposition we shall state a corollary which is more directly applicable to the study of the J- and K- methods. We precede this by a helpful remark.

3.3. Remark. A simple analysis of $K(t,x)$ shows that

$$\min(1,1/t)K(t,x) = \begin{cases} K(t,x) & \text{if } t \le 1 \\ \inf\{1/t\|x_0\|_{X_0} + \|x_1\|_{X_1} : x = x_0 + x_1\} & \text{if } t > 1. \end{cases}$$

Therefore, $\min(1,1/t)K(t,x) \longrightarrow 0$ as $t \to 0$ or $t \to \infty$ in the following cases: (i) $x \in C\ell(\text{im } j)$, (ii) $x \in K(\theta,q,\overline{X})$, $1 \le q \le \infty$. Case (i) follows from Proposition 1.2 (iii), since for $x \in \Delta\overline{X}$,

$J(1,x) = \|x\|$. Case (ii) follows from the definition of $K(\theta,q,\bar{X})$.

3.4. <u>Corollary</u>. Let $x \in \Sigma\bar{X}$ and let $\{u_k\}$ be as in Proposition 3.2. Let $z_n = \sum_{-n+1}^{n} u_k$. (1) If $x \in K(\theta,q,\bar{X})$ for $1 \le q < \infty$, then the sequence $\{z_n\}_{n=1}^{\infty}$ is a Cauchy sequence in $J(\theta,q,\bar{X})$. (2) If $x \in K(\theta,\infty,\bar{X})$, then the sequence $\{z_n\}_{n=1}^{\infty}$ is uniformly bounded in $J(\theta,\infty,\bar{X})$. (3) $\{j(z_n)\}$ converges to x in $\Sigma\bar{X}$.

<u>Proof of Corollary</u>: (1) We note that for $n < m$

$$z_m - z_n = \sum_{-m+1}^{-n} u_k + \sum_{n+1}^{m} u_k.$$

Now by Proposition 3.2 we have

$$\left[\sum_{n+1}^{m} (2^{-\theta k} J(2^k,u_k))^q\right]^{1/q} \le 4\left[\sum_{n+1}^{m} (2^{-\theta k} K(2^k,x))^q\right]^{1/q} \quad \text{and}$$

$$\left[\sum_{-m+1}^{-n} (2^{-\theta k} J(2^k,u_k))^q\right]^{1/q} \le 4\left[\sum_{-m+1}^{-n} (2^{-\theta k} K(2^k,x))^q\right]^{1/q}.$$

However, since $\left[\sum_{-\infty}^{\infty} (2^{-\theta k} K(2^k,x))^q\right]^{1/q}$ is a convergent series, the remainder tends to 0, which proves that

$$\|z_n - z_m\| \to 0.$$

(2) By definition,

$$\|z_n\|_{J(\theta,\infty,\bar{X})} \le \sup(2^{-\theta k} J(2^k,u_k)).$$

Hence, by Proposition 3.2,

$$\|z_n\|_{J(\theta,\infty,\bar{X})} \le 4\sup(2^{-\theta k}K(2^k,x)) \le 4\|x\|_{K(\theta,\infty,\bar{X})}.$$

(3) From the above remark and the proposition, it follows that $\|x-j(z_n)\|_{\Sigma\bar{X}} \to 0$ as $n \to \infty$. □

Proof of Proposition 3.2: By definition of $K(t,x)$ there exist $x_{0,k} \in X_0$ and $x_{1,k} \in X_1$ such that $x = \sigma_0(x_{0,k}) + \sigma_1(x_{1,k})$ and

$$\|x_{0,k}\|_{X_0} + 2^k\|x_{1,k}\|_{X_1} \le 4/3\ K(2^k,x).$$

Now we define

$$y_{0,k} = x_{0,k} - x_{0,k-1} \quad \text{and} \quad y_{1,k} = x_{1,k-1} - x_{1,k}.$$

Then

$$\sigma(y_{0,k}) - \sigma_1(y_{1,k}) = \sigma_0(x_{0,k}) - \sigma_0(x_{0,k-1}) - \sigma_1(x_{1,k-1}) + \sigma_1(x_{1,k})$$

$$= x - x = 0,$$

so there exists $u_k \in \Delta\bar{X}$ such that

$$\delta_0(u_k) = y_{0,k} \quad \text{and} \quad \delta_1(u_k) = y_{1,k}.$$

Property (i) of $\{u_k\}$ is verified by the following calculation:

$$\begin{aligned} J(2^k,u_k) &= \max(\|y_{0,k}\|,\ 2^k\|y_{1,k}\|) \\ &= \max(\|x_{0,k} - x_{0,k-1}\|,\ 2^k\|x_{1,k-1} - x_{1,k}\|) \\ &\le \|x_{0,k}\| + 2^k\|x_{1,k}\| + 2(\|x_{0,k-1}\| + 2^{k-1}\|x_{1,k}\|) \\ &\le 4/3\ K(2^k,x) + 2\cdot 4/3\ K(2^{k-1},x) \le 4K(2^k,x). \end{aligned}$$

Moreover, by construction we have

$$x - j(\sum_{n+1}^{m} u_k) = x - \sum_{n+1}^{m} \sigma_0\delta_0(u_k) = x - \sum_{n+1}^{m} \sigma_0(y_{0,k})$$
$$= x - \sum_{n+1}^{m} (\sigma_0(x_{0,k}) - \sigma_0(x_{0,k-1}))$$
$$= x - \sigma_0(x_{0,m}) + \sigma_0(x_{0,n})$$
$$= \sigma_1(x_{1,m}) + \sigma_0(x_{0,n}).$$

Therefore,

$$\|x - j(\sum_{n+1}^{m} u_k)\| \le \|x_{1,m}\|_{X_1} + \|x_{0,n}\|_{X_0}$$
$$\le \min(1,2^{-m})K(2^m,x) + \min(1,2^{-n})K(2^n,x). \qquad \square$$

We are now in a position to be able to state and prove the following propositions, which in reality comprise half of the equivalence theorem.

3.5. <u>Proposition</u>. Let $\overline{X}$ be a doolittle diagram and let θ and q be given, $0 < \theta < 1$, $1 \le q < \infty$. Let $\varphi: J(\theta,q,\overline{X}) \to K(\theta,q,\overline{X})$ be the canonical map defined in Proposition 3.1. Then φ is surjective.

<u>Proof</u>: Let $x \in K(\theta,q,\overline{X})$. We know that the sequence $\{z_n\}_{n=1}^{\infty}$ given by Corollary 3.4 is a Cauchy sequence, and since φ is continuous, $\{\varphi(z_n)\}_{n=1}^{\infty}$ is a Cauchy sequence. Hence, it suffices to prove that $\|\varphi(z_n)-x\|_{K(\theta,q,\overline{X})} \to 0$. This follows directly since $\|\varphi(z_n)-x\|_{\Sigma\overline{X}} \to 0$. $\square$

3.6. <u>Proposition</u>. For $q=\infty$, the map $\varphi: J(\theta,\infty,\overline{X}) \to K(\theta,\infty,\overline{X})$ has closed range $K^0(\theta,\infty,\overline{X})$, where as in Theorem 2.1,

$$K^0(\theta,\infty,\overline{X}) = \{x \in K(\theta,\infty,\overline{X}) \mid t^{-\theta}K(t,x) \to 0 \text{ as } t \to 0 \text{ or } \infty\}.$$

Furthermore, the ball of radius 4 in $J(\theta,\infty,\overline{X})$ is dense in the unit ball of $K(\theta,\infty,\overline{X})$ with respect to the weak topology given by $\Delta\overline{X}'=(\Sigma\overline{X})'$.

Proof: Since $\varphi(\Delta\overline{X}) \subset K^0(\theta,\infty,\overline{X})$, it follows that $\varphi(J(\theta,\infty,\overline{X})) \subset K^0(\theta,\infty,\overline{X})$. Now if $x \in K^0(\theta,\infty,\overline{X})$, one can easily see that $\|z_n-z_m\|_{J(\theta,\infty,\overline{X})} \to 0$, so $\{z_n\}$ is a Cauchy sequence in $J(\theta,\infty,\overline{X})$. This shows that $\mathrm{im}(\varphi) = K^0(\theta,\infty,\overline{X})$.

Finally, let $x \in K(\theta,\infty,\overline{X})$. The sequence $\{z_n\}$ has the property that $\varphi(z_n) \to x$ in $\Sigma\overline{X}$ and thus also converges weakly with respect to $\Delta\overline{X}'$. Furthermore, we have

$$\|z_n\|_{J(\theta,\infty,\overline{X})} \le 4\|x\|_{K(\theta,\infty,\overline{X})},$$

as we showed in the proof of Corollary 3.4. □

Finally, we may prove the equivalence theorem.

3.7. Theorem. Let $\overline{X}$ be a doolittle diagram. Let θ and q be given, $0 < \theta < 1$ and $1 \le q \le \infty$. Then

(i) $J(\theta,q,\overline{X}) \approx K(\theta,q,\overline{X})$ (isomorphic up to equivalence of norms) for $1 \le q < \infty$, and (ii) $J(\theta,\infty,\overline{X}) \approx K^0(\theta,\infty,\overline{X})$.

Proof: If $1 \le q < \infty$, we know that the map $\varphi:J(\theta,q,\overline{X}) \to K(\theta,q,\overline{X})$ is surjective, so it remains to prove that φ is also injective. To do this we shall prove that the map $\sigma\circ\varphi:J(\theta,q,\overline{X}) \to \Sigma\overline{X}$ is injective. Now $\sigma\circ\varphi$ is injective if $(\sigma\circ\varphi)':(\Sigma\overline{X})' \to J(\theta,q,\overline{X})'$ is surjective in the sense that $\mathrm{im}((\sigma\circ\varphi)')$ is weak* dense in $J(\theta,q,\overline{X})'$. However, since $(\Sigma\overline{X})'=\Delta\overline{X}'$ and $J(\theta,q,\overline{X})' = K(\theta,q',\overline{X}')$ by 2.2, we may consider $(\sigma\circ\varphi)'$ (by abuse of notation) to be

$$\varphi\circ\delta:\Delta\overline{X}' \to K(\theta,q',\overline{X}').$$

Now if $1 < q \leq \infty$, so that $1 \leq q' < \infty$, then it follows from the definition of $J(\theta,q',\overline{X}')$ and Proposition 3.5 that $\mathrm{im}(\varphi\circ\delta)$ is even norm dense in $K(\theta,q',\overline{X}')$. If $q=1$, so $q'=\infty$, then it follows from Proposition 3.6 that $\mathrm{im}(\varphi\circ\delta)$ is dense with respect to a somewhat stronger topology than the weak* topology, and this concludes the proof of (i). We have also proved that φ is injective if $q=\infty$, so it follows from Proposition 3.6 that $J(\theta,\infty,\overline{X}) \approx K^0(\theta,\infty,\overline{X})$. □

A consequence of Propositions 3.5 and 3.6 is that the spaces $K(\theta,q,\overline{X})$, $1 \leq q < \infty$, and the space $K^0(\theta,\infty,\overline{X})$ are all regular in the sense that $\Delta\overline{X}$ is dense in them. Therefore, we may state the following corollary, which completes our duality theorem 2.1. The proof is left to the reader.

3.8. <u>Corollary</u>. Let $0 < \theta < 1$. The dual of $K(\theta,q,\overline{X})$ is (isometrically) isomorphic to $J(\theta,q',\overline{X}')$ for $1 \leq q < \infty$, and the dual of $K(\theta,\infty,\overline{X})$ is the space $J(\theta,1,\overline{X}')$.

3.9. <u>Remark</u>. Although we shall not work out the details, the "reiteration theorem" for the real method also holds in our setting, since it is a consequence of the equivalence theorem. In particular, if $\overline{A}$ and $\overline{X}$ are doolittle diagrams with $X_i = J(\theta_i,q,\overline{A})$, $0 \leq \theta_i \leq 1$, $\theta_0 \neq \theta_1$ then given η such that $0 < \eta < 1$, we have

$$J(\eta,q,\overline{X}) = J(\theta,q,\overline{A}),$$

where $\theta = (1-\eta)\theta_0 + \eta\theta_1$.

CHAPTER III

THE COMPLEX METHOD

1. The General Duality Theorem.

The classical complex method of interpolation theory was given in the present form in a famous paper by Calderón [4]. In contrast to our method in Chapter II of extending the real method to the $\overline{\mathfrak{B}}$-setting, we shall in the present Chapter see that the most obvious extension of the complex method to the category of doolittle diagrams works and that the Calderón duality theorem extends with no difficulty to the generalized setting. In this section we lay the categorical foundations necessary to obtain this duality theorem.

We recall that in Chapter I we introduced the operation $\overline{X} \longmapsto \overline{X}/K\overline{X}$ of making a doolittle diagram into a Banach couple and the operation $\overline{X} \longmapsto \overline{X}^0$ of "regularizing" $\overline{X}$. We noted in I.2.2 that the operations are commutative. We shall use here the notation introduced there. In particular, for the doolittle diagram

$$\overline{X} = \begin{array}{ccc} \Delta\overline{X} & \xrightarrow{\delta_0} & X_0 \\ \delta_1 \big\downarrow & \searrow^{j} & \big\downarrow \sigma_0 \\ X_1 & \xrightarrow[\sigma_1]{} & \Sigma\overline{X} \end{array} ,$$

we let $K_i\overline{X} = \ker \sigma_i$, $X_i^0 = \mathcal{C}\ell(\mathrm{im}(\delta_i))$, $K\overline{X} = \ker j$, and we let $\overline{K\overline{X}}$ denote the doolittle diagram

$$\overline{K\overline{X}} = \quad \begin{array}{ccc} K\overline{X} & \longrightarrow & K_0\overline{X} \\ \downarrow & & \downarrow \\ K_1\overline{X} & \longrightarrow & 0\ . \end{array}$$

We begin with the following simple proposition concerning the dual $\overline{X}'$ of $\overline{X}$, where as in I.1.6,

$$\overline{X}' = \quad \begin{array}{ccc} (\Sigma\overline{X})' = \Delta\overline{X}' & \xrightarrow{\ \sigma_0'\ } & X_0' \\ {\scriptstyle \sigma_1'}\downarrow & {\scriptstyle j'} & \downarrow{\scriptstyle \delta_0'} \\ X_1' & \xrightarrow[\ \delta_1'\ ]{} & \Sigma\overline{X}' = (\Delta\overline{X})'\ . \end{array}$$

1.1. Proposition. For any $\overline{X} \in \mathfrak{B}$ we have

(i) $\ker(\delta_i') = \mathrm{im}(\delta_i)^{\perp}$, so $(X_i^0)' = X_i'/K_i\overline{X}'$,

(ii) $\ker(j') = \mathrm{im}(j)^{\perp}$, so $(\Sigma\overline{X}^0)' = \Delta\overline{X}'/K\overline{X}'$,

(iii) $\mathrm{im}(\sigma_i') \subset \ker(\sigma_i)^{\perp}$, so $(X_i')^0 = ((X_i/K_i\overline{X})')^0$, and

(iv) $\Sigma(\overline{X}')^0 = ((\Delta\overline{X}/K\overline{X})')^0$.

Proof: We interpret $\Delta\overline{X}$ as a subspace of $X_0 \pi X_1$ and $\Sigma\overline{X}$ as a quotient $X_0 \amalg X_1/\Delta\overline{X}^-$ where $\Delta\overline{X}^- = \{(x_0,x_1) \mid (x_0 - x_1) \in \Delta\overline{X}\}$ (see I.1.2). Then $\Delta\overline{X}' = (\Sigma\overline{X})' = (\Delta\overline{X}^-)^{\perp} \subset X_0' \pi X_1'$ and $\Sigma\overline{X}' = (\Delta\overline{X})' = (X_0' \amalg X_1')/(\Delta\overline{X})^{\perp}$. We observe that $(\Delta\overline{X})^{\perp} \subset X_0' \amalg X_1'$ and

$$\begin{aligned} (\Delta\overline{X})^{\perp} &= \{(x_0',x_1') \mid \langle x_0',x_0\rangle + \langle x_1',x_1\rangle = 0 \text{ for all } (x_0,x_1) \in \Delta\overline{X}\} \\ &= \{(x_0',x_1') \mid \langle x_0',x_0\rangle + \langle -x_1',x_1\rangle = 0 \text{ for all } (x_0,x_1) \in \Delta\overline{X}^-\}, \end{aligned}$$

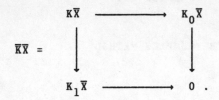

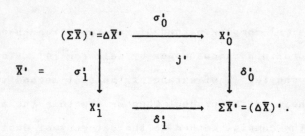

i.e. that $(\Delta\overline{X})^{\perp} = ((\Delta\overline{X}^{-})^{\perp})^{-}$.

To prove the proposition we first notice that in view of the above descriptions, (i), (ii), and (iii) are standard facts from the theory of Banach spaces. Therefore, it remains only to prove (iv). Towards this we simply observe that $\mathrm{im}(j') \subset (\ker j)^{\perp}$ and since $\Sigma(\overline{X}')^{0}$ is the norm closure of $\mathrm{im}(j')$, it does not matter if the closure is taken in $\Sigma\overline{X}'$ or in its subspace $(\ker j)^{\perp} = (\Delta\overline{X}/\ker j)'$. □

Let us denote by $(\overline{K}\overline{X})^{\perp}$ the dual diagram of $\overline{X}/\overline{K}\overline{X}$:

$$
(\overline{K}\overline{X})^{\perp} = \quad
\begin{array}{ccc}
(\Sigma\overline{X})' & \xrightarrow{\ \sigma_0'\ } & (K_0\overline{X})^{\perp} = (X_0/K_0\overline{X})' \\
\Big\downarrow \sigma_1' & & \Big\downarrow \delta_0' \\
(X_1/K\overline{X})' = (K_1\overline{X})^{\perp} & \xrightarrow[\ \delta_1'\ ]{} & (K\overline{X})^{\perp} = (\Delta\overline{X}/K\overline{X})'.
\end{array}
$$

Then from the above proposition we have the following corollary.

1.2. <u>Corollary</u>. Let $\overline{X} \in \overline{\mathfrak{B}}$. Then

(i) $(\overline{X}^{0})' = \overline{X}'/\overline{K}\overline{X}'$ and

(ii) $(\overline{X}^{0}/\overline{K}\overline{X})' = (\overline{K}\overline{X})^{\perp}/\overline{K}\overline{X}'$.

<u>Proof</u>: The statement (i) is completely obvious from Proposition 1.1. To prove (ii), we use (i) to get

$$(\overline{X}^{0}/\overline{K}\overline{X})' = (\overline{K}\overline{X})^{\perp}/\overline{K}((\overline{K}\overline{X})^{\perp}).$$

Moreover, in view of the fact that $\ker(\delta_i') = \mathrm{im}(\delta_i)^{\perp} \subset (K_i\overline{X})^{\perp}$, we can see that $\overline{K}((\overline{K}\overline{X})^{\perp}) = \overline{K}\overline{X}'$, which gives the desired result. □

A more important fact is that we also have the following result.

1.3. <u>Proposition</u>. For any $\overline{X}\in\mathfrak{B}$,

$$((\overline{X}^0/\overline{K}\overline{X})')^0 = (\overline{X}')^0/\overline{K}\overline{X}'.$$

<u>Proof</u>: We examine the definitions of the two constructions. $(\overline{X}')^0/\overline{K}\overline{X}'$ is obtained by completing the following diagram:

$$\begin{array}{ccc}
(\Sigma\overline{X})'/K\overline{X}' & \xrightarrow{\sigma_0'} & \mathcal{C}\ell(\mathrm{im}(\sigma_0'))\subset X_0'/K_0\overline{X}' \\
\sigma_1' \Big\downarrow & & \\
X_1'/K_1\overline{X}'\supset\mathcal{C}\ell(\mathrm{im}(\sigma_1')) & & ;
\end{array}$$

$((\overline{X}^0/\overline{K}\overline{X})')^0 = ((\overline{K}\overline{X})^\perp/\overline{K}\overline{X}')^0$ is obtained by completing the diagram:

$$\begin{array}{ccc}
(\Sigma\overline{X})'/K\overline{X}' & \xrightarrow{\sigma_0'} & \mathcal{C}\ell(\mathrm{im}(\sigma_0'))\subset(K_0\overline{X})^\perp/K_0\overline{X}' \\
\sigma_1' \Big\downarrow & & \\
(K_1\overline{X})^\perp/K_1\overline{X}'\supset\mathcal{C}\ell(\mathrm{im}(\sigma_1')) & & .
\end{array}$$

Moreover, we have $(K_i\overline{X})^\perp/K_i\overline{X}'\subset X_i/K_i\overline{X}'$, and since it does not matter in which space we take the closure of σ_i', the result follows. □

Using the above proposition, we can now prove the following "General Duality Theorem".

1.4. <u>Theorem</u>. Let F and G be interpolation functors (methods) defined on the category $\overline{\mathfrak{B}\mathfrak{C}}$ of Banach couples such that for every $\overline{X}\in\overline{\mathfrak{B}\mathfrak{C}}$, $F\overline{X} = F\overline{X}^0$ and $G\overline{X} = G\overline{X}^0$, and for every regular couple $\overline{X}$, $(F\overline{X})' = G\overline{X}'$. Then if we define $\tilde{F}$ and $\tilde{G}$ on $\mathfrak{B}$ by

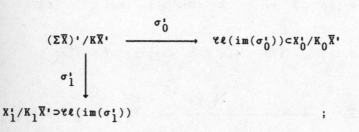

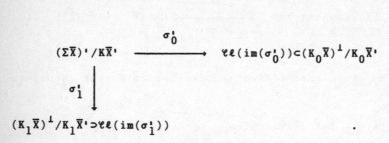

$$\tilde{F}\bar{X} = F(\bar{X}/\bar{K}\bar{X}),\ \tilde{G}\bar{X} = G(\bar{X}/\bar{K}\bar{X}),$$

it follows that for all $\bar{X}\in\mathfrak{B}$,

$$(\tilde{F}\bar{X})' = \tilde{G}\bar{X}'.$$

<u>Proof</u>: By assumption we have

$$\tilde{F}\bar{X} = F(\bar{X}/\bar{K}\bar{X}) = F((\bar{X}/\bar{K}\bar{X})^0) = F(\bar{X}^0/\bar{K}\bar{X})$$

and, hence,

$$(\tilde{F}\bar{X})' = (F(\bar{X}^0/\bar{K}\bar{X}))' = G((\bar{X}^0/\bar{K}\bar{X})').$$

But again by hypothesis,

$$G((\bar{X}^0/\bar{K}\bar{X})') = G(((\bar{X}^0/\bar{K}\bar{X})')^0).$$

By Proposition 1.3 and the assumptions, it follows that

$$G(((\bar{X}^0/\bar{K}\bar{X})')^0) = G((\bar{X}')^0/\bar{K}\bar{X}') = G((\bar{X}'/\bar{K}\bar{X}')^0)$$
$$= G(\bar{X}'/\bar{K}\bar{X}') = \tilde{G}(\bar{X}'),$$

which proves the result. □

1.5. <u>Remark</u>. The above result is actually an abstract duality theorem for functors extended from the category of Banach couples to the category of doolittle diagrams. The assumption that F and G are interpolation functors is made for motivation only and is not used in the proof.

2. The Duality Theorem.

Our method will be to apply the preceding result (Theorem 1.4) to prove that if the complex methods are extended to the category $\overline{\mathfrak{B}}$ in a straightforward way, then the Calderón duality theorem extends to the general case.

We proceed now to define the two interpolation functors C_θ and C^θ following the classical manner. Let $\overline{X}\in\overline{\mathfrak{B}}$. Then we define $A(S,\overline{X})$ to be the set of all $\Sigma\overline{X}$-valued functions which are continuous on the standard strip

$$S = \{z\in\mathbb{C} \mid 0 \le \operatorname{Re} z \le 1\},$$

analytic on its interior, and such that $f(k+it) \in \operatorname{im}(\sigma_k) = X_k/K_k\overline{X}$, $k=0,1$, and

$$\|f\| = \max(\sup_t \|f(it)\|_{X_0/K_0\overline{X}}, \sup_t \|f(1+it)\|_{X_1/K_1\overline{X}}) < \infty.$$

2.1. Definition. For $0 < \theta < 1$, we define

$$C_\theta\overline{X} = A(S,\overline{X})/I(\theta),$$

where $I(\theta) = \{f\in A(S,\overline{X}) \mid f(\theta)=0\}$.

2.2. Remarks. We note that $C_\theta\overline{X}$ is isometrically isomorphic to the set of all $x\in\Sigma\overline{X}$ such that $x=f(\theta)$ for some $f \in A(S,\overline{X})$ endowed with the norm $\|x\|_\theta = \inf\{\|f\|_{A(S,\overline{X})} \mid f(\theta)=x\}$.

As in the classical case we have the following proposition.

2.3. <u>Proposition</u>. The function which to every doolittle diagram $\overline{X}$ associates the space $C_\theta\overline{X}$ is a functor from $\overline{\mathfrak{B}}$ to $\mathfrak{B}$ such that $C_\theta\overline{X} = C_\theta(\overline{X}/\overline{K}\overline{X})$. Moreover, for every $T = (T_0,T_1) \in L(\overline{X},\overline{Y})$ the map $\Sigma T:\Sigma\overline{X} \to \Sigma\overline{Y}$ determines a continuous map $T:C_\theta\overline{X} \to C_\theta\overline{Y}$ such that $\|T\| \le \|T_0\|^{1-\theta}\|T_1\|^\theta$.

To define $C^\theta\overline{X}$ we consider the space $\Lambda(S,\overline{X})$ of $\Sigma\overline{X}$-valued functions continuous on S and analytic in the interior of S such that

$$\|f\|_\Lambda = \max\Big(\sup_{s<t} 1/t-s\|f(it)-f(is)\|_{X_0/K_0\overline{X}},\ \sup_{s<t} 1/t-s\|f(1+it)-f(1+is)\|_{X_1/K_1\overline{X}}\Big) < \infty.$$

Observing that it is necessary to reduce $\Lambda(S,\overline{X})$ by the constant function in order to obtain a Banach space, we shall consider $\Lambda(S,\overline{X})$ to be so reduced. For $0 < \theta < 1$ we have a map d_θ from $\Lambda(S,\overline{X})$ to $\Sigma\overline{X}$ defined by $d_\theta(f) = f'(\theta)$.

2.4. <u>Definition</u>. For $0 < \theta < 1$ we define

$$C^\theta\overline{X} = \Lambda(S,\overline{X})/K(\theta),$$

where $K(\theta) = \ker(d_\theta)$.

2.5. <u>Remarks</u>. We can see that $C^\theta\overline{X}$ is isometrically isomorphic to the set of all $x \in \Sigma\overline{X}$ such that $x=g'(\theta)$ for some $g \in \Lambda(S,\overline{X})$ endowed with the norm

$$\|x\|^\theta = \inf\{\|g\|_{\Lambda(S,\overline{X})} \mid g'(\theta)=x\}.$$

Like $C_\theta\overline{X}$, $C^\theta\overline{X}$ can also be shown to be an "exact interpolation space of exponent θ". Moreover, it is clear from the definition that $C^\theta\overline{X} = C^\theta(\overline{X}/\overline{KX})$.

We can now prove the duality theorem.

2.6. <u>Theorem</u>. For any $\overline{X}\in\overline{\mathfrak{B}}$ we have

$$(C_\theta\overline{X})' = C^\theta\overline{X}'.$$

Our method of proof will be to prove that the functors $F=C_\theta$ and $G=C^\theta$ satisfy the assumptions of the General Duality Theorem. Since the classical duality theorem gives the result on regular Banach couples and since $C_\theta\overline{X} = C_\theta(\overline{X}/\overline{KX})$ and $C^\theta\overline{X} = C^\theta(\overline{X}/\overline{KX})$, it remains for us to prove that for any Banach couple $\overline{X}$, $C_\theta\overline{X} = C_\theta\overline{X}^0$ and $C^\theta\overline{X} = C^\theta\overline{X}^0$. These facts will be proved by showing that $A(S,\overline{X}) = A(S,\overline{X}^0)$ and $\Lambda(S,\overline{X}) = \Lambda(S,\overline{X}^0)$.

In order to prove the first fact, we shall first define $A_0(S,\overline{X})$ to be the subspace of $A(S,\overline{X})$ consisting of all f such that $f_k\in C_0(\mathbb{R},X)$, $k=0,1$, where

$$f_k(t) = f(k+it),$$

i.e. $f_k(t) \to 0$ as $|t| \to \infty$.

Then we have the following fact.

2.7. <u>Proposition</u>. $A_0(S,\overline{X})/I(\theta) = A(S,\overline{X})/I(\theta)$.

<u>Proof</u>: Let $f \in A(S,\overline{X})$. Then for any $\varepsilon>0$, the function

$$f_\varepsilon(z) = \exp(\varepsilon(z-\theta)^2)f(z)$$

is in $A_0(S,\overline{X})$, $\|f_\varepsilon\| \leq \exp(\varepsilon)\|f\|$, and $f_\varepsilon(\theta) = f(\theta)$ (in fact also $f'_\varepsilon(\theta) = f'(\theta)$). Therefore, the result follows. □

The fact that $C_\theta\overline{X} = C_\theta\overline{X}^0$ follows from the following lemma, which also shows that $\Delta\overline{X}$ is dense in $C_\theta\overline{X}$. A proof of the lemma may be found in Bergh-Löfström [2].

2.8. Lemma. (Calderón) Let $\overline{X}$ be a Banach couple. Then the space of all linear combinations of functions of the form

$$\{\exp(\varepsilon(z-r)^2)x \,|\, \varepsilon > 0,\ r \in \mathbb{R},\ x \in \Delta\overline{X}\}$$

is dense in $A_0(S,\overline{X})$.

Since $\Delta\overline{X}$ is not in general dense in $C^\theta\overline{X}$, the proof that $\Lambda(S,\overline{X}) = \Lambda(S,\overline{X}^0)$ is somewhat more involved.

2.9. Lemma. $\Lambda(S,\overline{X}) = \Lambda(S,\overline{X}^0)$.

Proof: Let $f \in \Lambda(S,\overline{X})$. We want to show that $f \in \Lambda(S,\overline{X}^0)$, i.e. that $f(k+it) \in X_k^0$, for $k=0,1$. For this it suffices to show that $g_k(t) = g(k+it) \in X_k^0$, $k=0,1$, where

$$g(z) = \exp(z^2)f(z),$$

since then $\exp(-(k+it)^2)g(k+it) = f(k+it)$ is also in X_k^0. We shall use a Fourier transform argument to obtain this result. Thus, we define for $0 \leq x \leq 1$,

$$\hat{g}(x,\xi) = \int_{-\infty}^{\infty} g(x+it)\exp(\xi(x+it))dt.$$

Since g is analytic, it follows from Cauchy's theorem that $\hat{g}(x,\xi)$

is independent of x. Now $\hat{g}(0,\xi) = \int_{-\infty}^{\infty} g(it)e^{i\xi t}dt$, and since $g_0 \in L^1(\mathbb{R},X_0)$, it follows that $\hat{g}(0,\xi) \in X_0$. Likewise,

$$\begin{aligned}\hat{g}(1,\xi) &= \int_{-\infty}^{\infty} g(1+it)e^{\xi}e^{\xi it}\,dt \\ &= e^{\xi}\int_{-\infty}^{\infty} g(1+it)e^{i\xi t}\,dt,\end{aligned}$$

and since $g_1 \in L^1(\mathbb{R},X_1)$, it follows that $\hat{g}(1,\xi) \in X_1$. However, as observed above, $\hat{g}(x,\xi)$ is independent of x, so $\hat{g}(x,\xi) \in X_0 \cap X_1 = \Delta\overline{X}$, i.e. $\hat{g} \in C_0(\mathbb{R},\Delta\overline{X})$.

Now we shall prove that $g(it) \in X_0^0$. (The same argument will prove that $g(1+it) \in X_1^0$.) We observe that the function

$$g_\delta(it) = \int_{-\infty}^{\infty} e^{-\delta\xi^2}\hat{g}(\xi)e^{i\xi t}\,dt$$

belongs to $\Delta\overline{X}$ for every $\delta>0$. Hence, it suffices to prove that

$$\|g_\delta(it)-g(it)\|_{X_0} \to 0$$

as $\delta \to 0$. This follows from standard calculations which will be omitted, and, hence, our proof is completed. □

2.10. Remarks. (1) C_θ is what is known as the standard complex method of interpolation. For certain doolittle diagrams $\overline{X}$, $C_\theta\overline{X}$ is equivalent to $C^\theta\overline{X}$ (see [2]). (2) It was proved by Peetre [20] that it is possible to replace the space $\Lambda(S,\overline{X})$ in the definition of $C^\theta(\overline{X})$ by the space $H_\infty(S,\overline{X})$ which is defined as the space of bounded, analytic $\Sigma\overline{X}$-valued functions on S such that if

$f \in H_\infty(S,\overline{X})$, then f has boundary values f_0 and f_1 such that $f_0 \varepsilon\ L(L^1(\mathbb{R}),X_0)$ and $f_1 \varepsilon\ L(L^1(\mathbb{R}),X_1)$. We shall return to this construction in Chapter IX.

PART II

CHAPTER IV

CATEGORICAL NOTIONS

1. Categories of Doolittle Diagrams.

We shall take the basic concepts of category, functor, natural transformation, and adjointness as not entirely unfamiliar to the reader, although we shall review them as they occur in special examples. Thus, we start by defining doolittle diagrams in an arbitrary category $\mathcal{C}$.

1.1. Definitions. 1. A diagram

$$D = \begin{array}{ccc} P & \xrightarrow{\ u\ } & X \\ {\scriptstyle v}\downarrow & & \downarrow{\scriptstyle f} \\ Y & \xrightarrow[\ g\ ]{} & Q \end{array}$$

in $\mathcal{C}$ is called a pushout if it is commutative and if Q is "universal" in the sense that if $h:X \to R$ and $k:Y \to R$ are maps such that $h \circ u = k \circ v$, then there exists a unique map $\varphi:Q \to R$ such that $\varphi \circ f = h$ and $\varphi \circ g = k$. 2. D is said to be a pullback if P satisfies the dual (in the categorical sense of reversing arrows) universal condition. 3. Finally, D is a doolittle diagram if it is both a pullback and a pushout.

Morphisms of doolittle diagrams are defined in the obvious way: if D and D' are given by

$$D = \begin{array}{ccc} P & \xrightarrow{u} & X \\ {\scriptstyle v}\downarrow & & \downarrow{\scriptstyle f} \\ Y & \xrightarrow[g]{} & Q \end{array} \quad \text{and} \quad D' = \begin{array}{ccc} P' & \xrightarrow{u'} & X' \\ {\scriptstyle v'}\downarrow & & \downarrow{\scriptstyle f'} \\ Y' & \xrightarrow[g']{} & Q' \end{array} ,$$

then a pair (s,t) is a morphism from D to D' if $s:X \to X'$ and $t:Y \to Y'$ are morphisms in $\mathcal{C}$ such that the following diagram is commutative:

$$\begin{array}{ccccccc} & & X & \xrightarrow{s} & X' & & \\ & \overset{u}{\nearrow} & & & & \overset{f'}{\searrow} & \\ P & & & & & & Q' \\ & \underset{v}{\searrow} & & & & \underset{g'}{\nearrow} & \\ & & Y & \xrightarrow[t]{} & Y' & & \end{array} .$$

We remark that by the property of the pushout the map from P to Q' factors through Q, in particular, that there exists $q:Q \to Q'$ such that $f' \circ s = q \circ f$ and $g' \circ t = q \circ g$. Likewise, there is a factorization through P'. Thus, a morphism from D to D' is determined by (s,t) but is defined on all objects of the diagram D.

We shall denote by $\overline{\mathcal{C}}$ the category having as objects all doolittle diagrams in $\mathcal{C}$ and morphisms as described above.

The following description of the set of morphisms from D to D', denoted $\overline{\mathcal{C}}(D,D')$, is merely a convenient restatement of the definition of morphism.

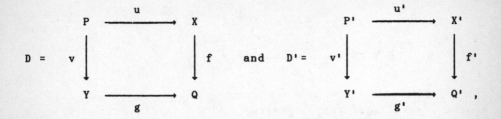

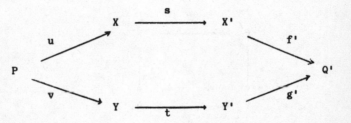

1.2. <u>Proposition</u>. $\overline{\mathcal{C}}(D,D')$ is the pullback (in the category of Sets)

of the diagram

$$
\begin{array}{ccc}
 & & \mathscr{C}(X,X') \\
 & & \big\downarrow \varphi \\
\mathscr{C}(Y,Y') & \xrightarrow[\psi]{} & \mathscr{C}(P,Q'),
\end{array}
$$

where $\varphi(s) = f'\circ s\circ u$ and $\psi(t) = g'\circ t\circ v$.

It will be convenient to have a notation for the transposed diagram of the diagram D.

1.3. Definition. If D is the doolittle diagram

$$
\begin{array}{ccc}
P & \xrightarrow{u} & X \\
{\scriptstyle v}\big\downarrow & & \big\downarrow{\scriptstyle f} \\
Y & \xrightarrow[g]{} & Q,
\end{array}
$$

then we shall write D^T to denote the diagram

$$
\begin{array}{ccc}
P & \xrightarrow{v} & Y \\
{\scriptstyle u}\big\downarrow & & \big\downarrow{\scriptstyle g} \\
X & \xrightarrow[f]{} & Q.
\end{array}
$$

Some of the additional properties which enable a category $\mathscr{C}$ to be a useful setting for analysis are summarized in the definition of a closed category.

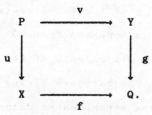

1.4. Definition. $\mathscr{C}$ is said to be a closed category if it is equipped with (i) a commutative and associative "tensor product" functor $\otimes_{\mathscr{C}}:\mathscr{C} \times \mathscr{C} \to \mathscr{C}$ and an "internal hom functor" $L_{\mathscr{C}}:\mathscr{C}^{op}\times \mathscr{C} \to \mathscr{C}$ (where $\mathscr{C}^{op}$, the "opposite category of $\mathscr{C}$", is a categorical notion for signaling contravariance of maps) such that $\otimes_{\mathscr{C}}Y$ is "strongly" left adjoint to $L_{\mathscr{C}}(Y,-)$, i.e.

$$L_{\mathscr{C}}(X\otimes_{\mathscr{C}}Y,Z) \cong L_{\mathscr{C}}(X,L_{\mathscr{C}}(Y,Z))$$

naturally in X, Y and Z, and (ii) an object I (the unit) such that $X\otimes_{\mathscr{C}}I \cong X$ and $L_{\mathscr{C}}(I,X) \cong X$ for all $X\in\mathscr{C}$. (A few behavioural properties, called coherence axioms, are required of $L_{\mathscr{C}}$ and $\otimes_{\mathscr{C}}$; see [18].)

Examples of closed categories include Banach spaces (of course), abelian groups, R-modules, and compactly generated Hausdorff spaces.

Categories $\mathscr{D}$ which are "based" on a closed category $\mathscr{C}$ - these are called <u>$\mathscr{C}$-categories</u> - in the sense that there is a "well-behaved" hom functor,

$$L:\mathscr{D}^{op} \times \mathscr{D} \to \mathscr{C},$$

are also rich in the sense of analysis, so it is a useful fact that $\overline{\mathscr{C}}$ has this property as well as being itself a closed category.

1.5. <u>Proposition</u>. Let $\mathscr{C}$ be a closed category with pushouts and pullbacks. Then $\overline{\mathscr{C}}$ is both a $\mathscr{C}$-category and a closed category.

<u>Proof</u>: If D and D' are two doolittle diagrams, we can define a $\mathscr{C}$-valued hom functor by taking the pullback L(D,D') (in $\mathscr{C}$) of the following diagram

$$\begin{array}{ccc} & & L_{\mathscr{C}}(X,X') \\ & & \big\downarrow \\ L_{\mathscr{C}}(Y,Y') & \longrightarrow & L_{\mathscr{C}}(P,Q'). \end{array}$$

Then we take the pushout of the diagram

$$\begin{array}{ccc} L(D,D') & \longrightarrow & L_{\mathscr{C}}(X,X') \\ \big\downarrow & & \\ L_{\mathscr{C}}(Y,Y') & & \end{array}$$

and define $\overline{L}(D,D')$ to be this resulting doolittle diagram. $L(-,-)$ makes $\overline{\mathscr{C}}$ into a $\mathscr{C}$-category, while $\overline{L}(-,-) = L_{\overline{\mathscr{C}}}(-,-)$ gives us an internal hom functor for $\overline{\mathscr{C}}$.

The tensor product is defined similarly. We first define the $\mathscr{C}$-valued tensor product $D\otimes D'$ to be the pushout of the diagram

$$\begin{array}{ccc} P\otimes_{\mathscr{C}}P' & \longrightarrow & X\otimes_{\mathscr{C}}X' \\ \big\downarrow & & \\ Y\otimes_{\mathscr{C}}Y' & & \end{array}$$

and then define $D\overline{\otimes}D'$ to be the doolittle diagram which results by taking the pullback of

$$\begin{array}{ccc} & & X\otimes_{\mathscr{C}}X' \\ & & \big\downarrow \\ Y\otimes_{\mathscr{C}}Y' & \longrightarrow & D\otimes D'. \end{array}$$

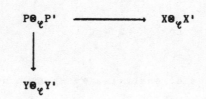

Finally, the constant diagram $\overline{I}$ with I (the unit of $\mathscr{C}$) at all vertices provides a unit for the tensor product $\overline{\mathscr{C}}$.

It is a simple verification that all the required relations are satisfied so that $\overline{\mathscr{C}}$ is indeed a closed category. □

We close this section by giving some relations between $\mathscr{C}$ and $\overline{\mathscr{C}}$. Clearly, there is an obvious diagonal embedding

$$J:\mathscr{C} \to \overline{\mathscr{C}}$$

defined (on objects) by

$$JX = \begin{array}{ccc} X & \xrightarrow{1} & X \\ {\scriptstyle 1}\downarrow & & \downarrow{\scriptstyle 1} \\ X & \xrightarrow[1]{} & X \end{array} .$$

That J has both a left and a right adjoint, namely the pushout and pullback operations, respectively, is expressed by the following proposition.

1.6. <u>Proposition</u>. Given $Z\in\mathscr{C}$ and $D\in\overline{\mathscr{C}}$, where

$$D = \begin{array}{ccc} P & \longrightarrow & X \\ \downarrow & & \downarrow \\ Y & \longrightarrow & Q \end{array} ,$$

we have

$$\mathscr{C}(Q,Z) = \overline{\mathscr{C}}(D,JZ) \quad \text{and}$$
$$\mathscr{C}(Z,P) = \overline{\mathscr{C}}(JZ,D).$$

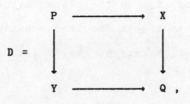

Proof: The proof is immediate from the definitions of pushouts and pullbacks. □

2. Doolittle Diagrams of Banach Spaces.

We now return to the category $\mathcal{C}=\mathcal{B}$, where $\mathcal{B}$ denotes the category of Banach spaces over the scalar field I (which indifferently denotes either the field of real numbers or the field of complex numbers) and norm decreasing linear maps. It is well known that $\mathcal{B}$ is a closed category having all set-indexed limits and colimits. Its internal hom is the space of all bounded linear maps from X to Y, denoted simply by L(X,Y) rather than by $L_{\mathcal{B}}(X,Y)$. Thus, the morphisms in the category $\mathcal{B}$, $\mathcal{B}(X,Y)$, form the closed unit ball of the space L(X,Y). The Banach space tensor product $-\otimes Y$ (i.e. the projective tensor product) is left adjoint to the hom functor L(Y,-), i.e.

$$L(X\otimes Y,\ Z) = L(X,\ L(Y,Z)).$$

Thus, $\otimes$ plays the role of $\otimes_{\mathcal{B}}$ in the closed structure of $\mathcal{B}$. Clearly, the scalar field is a unit for this symmetric tensor product. Products and coproducts for $\mathcal{B}$ may be easily described: if $X_\alpha \in \mathcal{B}$, $\alpha \in A$, then

$$\Pi X_\alpha = \{(x_\alpha) \in \times X_\alpha \mid \sup \|x_\alpha\| < \infty\}$$

with the sup norm, and

$$\amalg X_\alpha = \{(x_\alpha) \in \times X_\alpha \mid x_\alpha = 0 \quad \text{for all but finitely many} \quad \alpha\}$$

with $\|(x_\alpha)\|_{\amalg X_\alpha} = \sum_{\alpha \in A} |x_\alpha|$.

Since pullbacks and pushouts exist in $\mathfrak{B}$, we know from the previous section that $\overline{\mathfrak{B}}$, the category of doolittle diagrams of Banach spaces, is both a $\mathfrak{B}$-category and a closed category in its own right. Thus, we have $\mathfrak{B}$-valued and $\overline{\mathfrak{B}}$-valued hom functors,

$$L(\overline{X},\overline{Y}) \in \mathfrak{B}, \ \overline{L}(\overline{X},\overline{Y}) \in \overline{\mathfrak{B}}.$$

(We shall abbreviate $L(\overline{X},\overline{X})$ by $L\overline{X}$.) We also have $\mathfrak{B}$-valued and $\overline{\mathfrak{B}}$-valued tensor products

$$\overline{X}\otimes\overline{Y} \in \mathfrak{B}, \ \overline{X}\overline{\otimes}\overline{Y} \in \overline{\mathfrak{B}}.$$

The strong adjointness of $\overline{L}(\overline{Y},-)$ and $\overline{\otimes}\overline{Y}$ implies that there is a natural isomorphism $L(\overline{X}\overline{\otimes}\overline{Y}, \overline{Z}) = L(\overline{X},\overline{L}(\overline{Y},\overline{Z}))$. We shall obtain a related and more useful adjointness below.

The typical element $\overline{X}$ of $\overline{\mathfrak{B}}$ will be a diagram labelled in the following way:

$$\begin{array}{ccc} \Delta\overline{X} & \xrightarrow{\ \delta_0\ } & X_0 \\ {\scriptstyle \delta_1}\downarrow & & \downarrow{\scriptstyle \sigma_0} \\ X_1 & \xrightarrow[\ \sigma_1\ ]{} & \Sigma\overline{X} \end{array} .$$

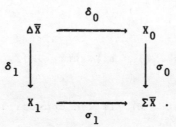

It is obvious from this notation that Δ and Σ are functors from $\overline{\mathfrak{B}}$ to $\mathfrak{B}$ assigning to each $\overline{X}\in\overline{\mathfrak{B}}$ the pullback and pushout, respectively, of the diagram. The following observations are completely obvious.

2.1. Proposition. $\Sigma\overline{X} = \overline{I}\overline{\otimes}\overline{X}$, $\Delta\overline{X} = L(\overline{I},\overline{X})$, and $\overline{X}\otimes JY = \Sigma\overline{X}\otimes Y$, where $J:\mathfrak{B} \to \overline{\mathfrak{B}}$ is the diagonal embedding.

We recall for the convenience of the reader that a concrete description of a doolittle diagram in $\mathfrak{B}$ was given in I.1.2 which showed that $\Delta\overline{X}$ is a closed subspace of the product space $X_0 \pi X_1$ and that the data $(\Delta\overline{X}, X_0, X_1)$ completely determines $\overline{X}$. It was also verified there that $\overline{\mathfrak{B}}$ is closed under dual diagrams, i.e. that $(\Sigma\overline{X})' = \Delta\overline{X}'$ and $(\Delta\overline{X})' = \Sigma\overline{X}'$, where $\overline{X}' = \overline{L}(\overline{X},\overline{I})$.

It follows from Proposition 1.6 that the functors Σ and Δ are, respectively, left and right adjoints to the diagonal embedding $J:\mathfrak{B} \to \overline{\mathfrak{B}}$, i.e. that

$$L(\Sigma\overline{X}, Y) \cong L(\overline{X}, JY) \quad \text{and}$$
$$L(Y, \Delta\overline{X}) \cong L(JY, \overline{X}).$$

Combining this result with the adjointness of $\overline{L}(\overline{Y},-)$ and $\overline{\otimes}\overline{Y}$, we obtain the following result.

2.2. Proposition. $-\overline{\otimes}\overline{Y}:\overline{\mathfrak{B}} \to \mathfrak{B}$ has a right adjoint $\overline{L}(\overline{Y}, J(-)):\mathfrak{B} \to \overline{\mathfrak{B}}$, i.e.

$$L(\overline{X}\otimes\overline{Y}, Z) = L(\overline{X}, \overline{L}(\overline{Y}, JZ)).$$

Proof: It is obvious that $-\otimes\overline{Y}$ is the composition of the functors:

$$\overline{\mathfrak{B}} \xrightarrow{\ -\overline{\otimes}\overline{Y}\ } \overline{\mathfrak{B}} \xrightarrow{\ \Sigma\ } \mathfrak{B}.$$

But Σ has J for its right adjoint, and $-\overline{\otimes}\overline{Y}$ has $\overline{L}(\overline{Y},-)$ for its right adjoint. It is a standard categorical fact that the composition of right adjoints is again a right adjoint. □

2.3. Corollary. $(\overline{X}\otimes\overline{Y})' = L(\overline{X},\overline{Y}')$.

Proof: Let $Z=I$. □

The tensor product of doolittle diagrams will play an important role in our theory of interpolation. One of the frequently occurring maps in connection with the tensor product is the trace map.

2.4. Definition. Since $(\overline{X}\otimes\overline{X}')' = L(\overline{X}')$ by 2.3, we may define the trace map

$$\mathrm{Tr}:\overline{X}\otimes\overline{X}' \to I$$

by $\mathrm{Tr}(t) = \langle 1_{\overline{X}'},t\rangle$.

Clearly, $\mathrm{Tr} = (\mathrm{Tr}_0,\mathrm{Tr}_1)$ is given by the usual trace maps

$$\mathrm{Tr}_i: X_i\otimes X_i' \to I, \quad \text{where}$$
$$\mathrm{Tr}_i(\Sigma x_i\otimes x_i') = \Sigma\langle x_i',x_i\rangle.$$

In addition to the categories $\mathfrak{B}$ and $\overline{\mathfrak{B}}$, we shall occasionally consider also the category $\mathfrak{B}_\infty$, which is the category of Banach spaces with all bounded linear maps as morphisms, and its corresponding category of doolittle diagrams, $\overline{\mathfrak{B}}_\infty$. Isomorphisms in $\mathfrak{B}_\infty$, i.e. topological isomorphisms, are denoted by $\approx$, while isomorphisms in $\mathfrak{B}$ are simply denoted by $=$. We note that $\overline{\mathfrak{B}}_\infty$ has the same objects as $\overline{\mathfrak{B}}$ since the maps in the doolittle diagram are contractions. However, the morphisms of $\overline{\mathfrak{B}}_\infty$ need not be contractions, i.e. $T:\overline{X} \to \overline{Y}$ is a $\overline{\mathfrak{B}}_\infty$-morphism if there exists $\lambda>0$ such that λT is a $\overline{\mathfrak{B}}$-morphism. In Chapters I-III we implicitly worked in $\overline{\mathfrak{B}}_\infty$ since this avoided the need to "normalize" such spaces as $J(\theta,p,\overline{X})$. In the present part of the work we shall generally work in $\overline{\mathfrak{B}}$ while remarking that nearly everything works as well for $\overline{\mathfrak{B}}_\infty$.

In most important (non-trivial) doolittle diagrams $\overline{X}$ of $\overline{\mathfrak{B}}$, the map $j:\Delta\overline{X} \to \Sigma\overline{X}$ turns out to have norm equal to 1. This means that there exists a sequence $\{x_n\} \subset \Delta\overline{X}$ such that $\|x_n\|_{\Delta\overline{X}} = 1$ and $\|jx_n\|_{\Sigma\overline{X}} \geq 1 - 1/n$. In many cases there actually exists some element $u\in\Delta\overline{X}$ such that $\|u\|_{\Delta\overline{X}} = \|ju\|_{\Sigma\overline{X}} = 1$. We shall call such an element a <u>unit</u> and say that a doolittle diagram is <u>unital</u> if it has a unit. We observe that if $\overline{X}$ is unital, then there exists $u'\in(\Sigma\overline{X})'$ such that $\|u'\| = \|j'(u')\| = 1$, so $\overline{X}'$ is unital with (dual) unit u'. If $\overline{X}$ is not unital but j has norm 1, then $\overline{X}$ is said to have an <u>approximating unit</u>. We shall frequently assume that our diagrams are unital in order to simplify proofs. However, this assumption is merely a technical convenience, for our proofs would work equally well with an approximating unit.

Finally, we wish to state that in the interest of simplicity and aesthetics, we shall avoid any extra notation which we believe the context makes superfluous. For example, δ, σ, and j will not be indexed by their home $\overline{X}$, and often ΔT and ΣT for $T:\overline{X} \to \overline{Y}$ will simply be denoted by T.

3. <u>Limits, Colimits, and Morphisms</u>.

It is a well known categorical fact that if a category has products and sums, kernels and cokernels, then it has all set-indexed limits and colimits. We shall begin this section by verifying that $\overline{\mathfrak{B}}$, like $\mathfrak{B}$, has this property.

3.1. <u>Proposition</u>. Let $\overline{X}_j = (X_{0j}, X_{1j}, \Delta\overline{X}_j)$, $j \in J$, be a set of elements of $\overline{\mathfrak{B}}$. Then

(i) $\pi\Delta\overline{X}_j$ is a closed subspace of $(\pi X_{0j})\pi(\pi X_{1j})$, so $(\pi X_{0j}, \pi X_{1j}, \pi\Delta\overline{X}_j)$ determines a doolittle diagram, $\pi\overline{X}_j$.

(ii) $\amalg\Delta\overline{X}_j$ is a closed subspace of $(\amalg X_{0j})\pi(\amalg X_{1j})$, so $(\amalg X_{0j}, \amalg X_{1j}, \amalg\Delta\overline{X}_j)$ determines a doolittle diagram, $\amalg\overline{X}_j$.

(iii) The diagrams $\pi\overline{X}_j$ and $\amalg\overline{X}_j$ satisfy the universal properties of a product and a sum in $\overline{\mathfrak{B}}$.

Since the proof is a simple verification, it is left to the reader.

3.2. <u>Proposition</u>. Let $T = (T_0, T_1): \overline{X} \to \overline{Y}$ be a morphism in $\overline{\mathfrak{B}}$. Then T has both a kernel and a cokernel satisfying the standard universal properties.

<u>Proof</u>: We first consider $\ker T_0 \subset X_0$ and $\ker T_1 \subset X_1$ and define $\Delta(\ker T)$ to be the pullback of the diagram

$$\begin{array}{ccc} & & \ker T_0 \\ & & \downarrow \\ \ker T_1 & \longrightarrow & \Sigma\overline{X} \end{array} .$$

In concrete terms $\Delta(\ker T) = (\ker T_0 \times \ker T_1) \cap \Delta\overline{X}$. Now we define $\Sigma(\ker T)$ to the pushout of the diagram

$$\begin{array}{ccc} \Delta(\ker T) & \longrightarrow & \ker T_0 \\ \downarrow & & \\ \ker T_1 & & \end{array} ,$$

which gives us a doolittle diagram $\overline{\ker T}$. We now shall show that $\overline{\ker T}$ is the kernel of T in the categorical sense, i.e. that if $S:\overline{Z} \to \overline{X}$ is such that $T \circ S = 0$, then there exists $R:\overline{Z} \to \overline{\ker T}$ such that $S = i \circ R$, where $i:\overline{\ker T} \to \overline{X}$ is the natural inclusion (in the sense that i_0 and i_1 are the inclusion maps). To see this we first observe that since $T_0 \circ S_0 = 0 = T_1 \circ S_1$, we have $\mathrm{im}(S_0) \subset \ker T_0$ and $\mathrm{im}(S_1) \subset \ker T_1$. Then since $\mathrm{im}(\Delta S) \subset \Delta\overline{X}$, by the description of $\Delta(\ker T)$, we see that S really defines a map $R:\overline{Z} \to \overline{\ker T}$ and this proves our assertion.

The cokernel, $\overline{\mathrm{cok}\, T}$, is defined in a dual way by considering the pushout of the diagram

$$\begin{array}{ccc} \Delta\overline{Y} & \longrightarrow & Y_0/\mathcal{C}\ell(\mathrm{im}\, T_0) \\ \downarrow & & \\ Y_1/\mathcal{C}\ell(\mathrm{im}\, T_1) & & \end{array}$$

and then taking the pullback of what we get. □

3.3. <u>Corollary</u>. $\overline{\mathfrak{B}}$ has all set-indexed limits and colimits.

3.4. <u>Remark</u>. The above constructions also show that $\overline{\mathfrak{B}}_\infty$ has all finite limits and colimits since 3.2 together with the existence of finite products and sums are sufficient to guarantee this fact. The main categorical difference between $\mathfrak{B}$ and $\mathfrak{B}_\infty$ is precisely that the former has infinite categorical products and sums while the latter does not.

We now shall examine the notions of monomorphism and epimorphism in $\overline{\mathfrak{B}}$. We first recall that in $\mathfrak{B}$ (categorical) monomorphisms, extremal monomorphisms, epimorphisms, and extremal epimorphisms are, respectively, one-to-one maps, isometric inclusions, maps with dense image, and quotient maps.

3.5. Definition. Let $T:\overline{X} \to \overline{Y}$ be a morphism. We shall say that (i) T is a monomorphism if whenever $S:\overline{Z} \to \overline{X}$ is such that $T \circ S = 0$, then $S=0$; (ii) T is an extremal monomorphism if T_0, T_1 are extremal monomorphisms and if ΣT is a monomorphism in $\mathfrak{B}$; (iii) T is an epimorphism if whenever $S:\overline{Y} \to \overline{Z}$ is such that $S \circ T = 0$, then $S=0$; (iv) T is an extremal epimorphism if T_0, T_1 are extremal epimorphisms and ΔT is an epimorphism in $\mathfrak{B}$.

It is straightforward to verify the following proposition.

3.6. Proposition. The following statements are equivalent:

(i) $T:\overline{X} \to \overline{Y}$ is a monomorphism (epimorphism);

(ii) T_0 and T_1 are monomorphisms (epimorphisms);

(iii) T_0, T_1, and ΔT are monomorphisms (T_0, T_1, and ΣT are epimorphisms);

(iv) $\overline{\ker T} = \overline{0}$ ($\overline{\operatorname{cok} T} = \overline{0}$).

We remark here that although $\overline{\ker T} = \overline{0}$ implies that T_0, T_1, and ΔT are monomorphisms, it does not follow that ΣT is a monomorphism as the following example illustrates: let $T = (1,1):\overline{X} \to \overline{Y}$, where

$$\overline{X} = \begin{array}{ccc} 0 & \longrightarrow & I \\ \big\downarrow & & \big\downarrow \\ I & \longrightarrow & I^2 \end{array} \quad \text{and} \quad \overline{Y} = \overline{I} = \begin{array}{ccc} I & \longrightarrow & I \\ \big\downarrow & & \big\downarrow \\ I & \longrightarrow & I \end{array} .$$

From this point on we shall not hesitate to abbreviate the words monomorphisms and epimorphism to mono and epi, respectively.

In analogy with the category of Banach spaces we shall make the following additional definition.

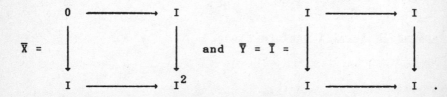

3.7. Definition. Let $T:\bar{X} \to \bar{Y}$ be a morphism. Then we say that (i) $\bar{X}$ is a subdiagram of $\bar{Y}$ if T is an extremal mono; (ii) $\bar{Y}$ is a quotient diagram of $\bar{X}$ if T is an extremal epi.

3.8. Proposition. If $\bar{X}$ is a subdiagram of $\bar{Y}$, then $\Delta\bar{X} = (X_0 \pi X_1) \cap \Delta\bar{Y}$.

Proof: Clearly, $\Delta\bar{X} \subset (X_0 \pi X_1) \cap \Delta\bar{Y}$. Now letting $T:\bar{X} \to \bar{Y}$ denote the inclusion map, if $(T_0x_0, T_1x_1) \in \Delta\bar{Y}$, we must show $x = (x_0, x_1) \in \Delta\bar{X}$, i.e. that $\sigma_0x_0 = \sigma_1x_1$. By definition of ΣT we have that $\Sigma T\sigma_ix_i = \sigma_iT_ix_i$, $i=0,1$. Hence, since $\sigma_0T_0x_0 = \sigma_1T_1x_1$, it follows that $\Sigma T\sigma_0x_0 = \Sigma T\sigma_1x_1$ and, therefore, since ΣT is mono, that $\sigma_0x_0 = \sigma_1x_1$.□

3.9. Remark. It is easy to see that the subdiagram of a Banach couple is again a couple. It will be shown in Chapter V that every doolittle diagram is the inductive limit of its finite dimensional subdiagrams and that every Banach couple is the inductive limit of its finite dimensional subcouples.

4. Functors and Natural Transformations.

Although we have taken the notions of functor and natural transformation as known, there are several aspects of these notions which must be discussed. First we note that since $\bar{\mathfrak{B}}$ is a $\mathfrak{B}$-category, we normally would be interested in functors $F:\bar{\mathfrak{B}} \to \mathfrak{B}$ or $F:\bar{\mathfrak{B}} \to \bar{\mathfrak{B}}$ which in acting on the morphisms preserve the structure with which the sets of morphisms are endowed, i.e. we would like the assignment

$$f \in L(\overline{X},\overline{Y}) \longmapsto Ff \in L(F\overline{X},F\overline{Y})$$

to be a norm-decreasing linear map of Banach spaces. (In the context of $\mathcal{C}$-categories, such a functor is called a "strong functor" or a "$\mathcal{C}$-functor".) In fact the functors we have discussed thus far, namely hom, tensor, Δ, Σ, K_i, K, $\overline{K}$, $(\ )^0$, and $(\ )/\overline{K}(\ )$ have all been strong functors in this sense, as will be all the other functors we shall meet. The notion of adjointness likewise should reflect the enriched structures of the hom sets, so we shall speak, say, of $F:\overline{\mathcal{B}} \to \mathcal{B}$ (or $F:\overline{\mathcal{B}} \to \overline{\mathcal{B}}$) being the (strong) left adjoint of $G:\mathcal{B} \to \overline{\mathcal{B}}$ $(G:\overline{\mathcal{B}} \to \overline{\mathcal{B}})$ when

$$L(F\overline{X},Y) = L(\overline{X},GY)\ (L(F\overline{X},\overline{Y}) = L(\overline{X},G\overline{Y})).$$

Any adjointness situation, say $F:\overline{\mathcal{B}} \to \mathcal{B}$ being left adjoint to $G:\mathcal{B} \to \overline{\mathcal{B}}$, gives rise to a natural transformation $\eta:\overline{X} \to GF\overline{X}$, the <u>unit map</u>, and $\varepsilon:FGY \to Y$, the <u>counit map</u>, corresponding, respectively, to the identity maps of $L(F\overline{X},F\overline{X})$ and $L(GY,GY)$ under the above isomorphism.

4.1. <u>Examples</u>. 1. We have already noted in Section 2 that the embedding $J:\mathcal{B} \to \overline{\mathcal{B}}$ has Σ and Δ as (strong) left and right adjoints. If we analyse the notion of the unit of the (Σ,J) adjunction and the counit of the (J,Δ) adjunction, we see that the first is just

$$(\sigma_0,\sigma_1):\overline{X} \to J\Sigma\overline{X}$$

and the second is just

$$(\delta_0,\delta_1):J\Delta\overline{X} \to \overline{X}.$$

This merely tells us that the σ's and δ's are natural in $\overline{X}$. 2. We saw in I.2 that every morphism $T:\overline{X} \to \overline{Y}$, where $\overline{X} \in \mathcal{B}$, $\overline{Y} \in \overline{\mathcal{B}}\mathcal{C}$, factors through $\overline{X}/\overline{K}\overline{X}$. This tells us that

$$L(\overline{X},\overline{Y}) = L(\overline{X}/\overline{K}\overline{X},\overline{Y}),$$

or that $(\ )/\overline{K}(\ )$ is the left adjoint of the inclusion functor from $\overline{\mathcal{B}}\mathcal{C}$ to $\overline{\mathcal{B}}$. Clearly the unit map is the canonical projection

$$\overline{X} \to \overline{X}/\overline{K}\overline{X}$$

and the counit map is just the identity on $\overline{Y}$. 3. A situation which has applications in our theory arises when we consider the category $\mathcal{B}^{\overline{\mathcal{B}}}$ of all functors from $\overline{\mathcal{B}}$ to $\mathcal{B}$ with natural transformations as morphisms and the functor

$$U:\mathcal{B}^{\overline{\mathcal{B}}} \to \mathcal{B}$$

defined by $UF = F(\overline{I})$. Then U has both left and right adjoints given, respectively, by $X \longmapsto L(\overline{I},-)\otimes X$ and $X \longmapsto L(L(-,\overline{I}),X)$. (One will see in Chapter VI that these are special cases of "Kan extensions", specifically, in the notation introduced there that $L(\overline{I},-)\otimes X = \mathrm{Lan}_X$ and $L(L(-,\overline{I}),X) = \mathrm{Ran}_X$.) We are particularly interested in the counit map that arises from the left adjoint of U and the unit map that arises from the right adjoint of U. Explicitly, these maps are

$$\varepsilon:L(\overline{I},\overline{X})\otimes F\overline{I} \to F\overline{X} \quad \text{and}$$
$$\eta:F\overline{X} \to L(L(\overline{X},\overline{I}),F\overline{I})$$

defined by $\varepsilon(T\otimes a) = FT(a)$ and $(\eta x)(S) = FS(x)$. We notice that both of these maps are $\mathfrak{B}$-maps and that both are types of evaluation maps. In deference, therefore, to the analysts among our readers we shall refer to ε as the evaluation map and to η as the coevaluation map.

In Example 3 above we see that for $F=\Delta$, ε is just the identity map since $L(\overline{I},\overline{X}) = \Delta\overline{X}$. If $F=\Sigma$, then η is the canonical inclusion since $L(\overline{X},\overline{I})' = (\Sigma\overline{X})'' = \Sigma\overline{X}''$. Partly for this reason, but mainly for the relation with interpolation functors to be discussed in the next section, we make the following definitions.

4.2. Definitions. Let $F:\overline{\mathfrak{B}} \to \mathfrak{B}$ be a functor. We say that (i) F is a Δ-functor if $\varepsilon:L(\overline{I},\overline{X})\otimes F\overline{I} \to F\overline{X}$ is an epi for all $\overline{X}\in\overline{\mathfrak{B}}$ and (ii) F is a Σ-functor if $\eta:F\overline{X} \to L(L(\overline{X},\overline{I}),F\overline{I})$ is a mono for all $\overline{X}\in\overline{\mathfrak{B}}$.

4.3. Remark. The maps ε, η have previously been studied in the category of Banach spaces. Mityagin and Švarc [19] called a functor $F:\mathfrak{B} \to \mathfrak{B}$ of type Σ when $\varepsilon:X\otimes FI \to FX$ is epi (since the functor $\Sigma_A = A\otimes-$ was the typical example) and this terminology was adopted also in [10] and [5]. We regret that it is necessary to change this terminology, but the reader will observe in the next section that this change is justifiable. Cigler, Losert, and Michor [5] say that $F:\mathfrak{B} \to \mathfrak{B}$ is total when $\eta:FX \to L(X',FI)$ is mono, but as far as we know the terminology has not had wide adoption.

Although we are frequently interested in specific natural transformations such as ε and η above, we are even more often considering the set of all natural transformations $NAT(F,G)$ between functors $F,G: \overline{\mathfrak{B}} \to \mathfrak{B}$ endowed with the structure of a Banach space. We recall in general that a natural transformation $t:F \to G$ is a collection of maps $\{t_{\overline{X}}:F\overline{X} \to G\overline{X} \mid \overline{X} \in \overline{\mathfrak{B}}\}$ which are "natural in $\overline{X}$" in the sense that given another doolittle diagram $\overline{Y}$ and a map $S:\overline{X} \to \overline{Y}$, the maps $t_{\overline{X}}$ and $t_{\overline{Y}}$ "fit together", i.e. that the following diagram is commutative:

$$\begin{array}{ccc} F\overline{X} & \xrightarrow{\;t_{\overline{X}}\;} & G\overline{X} \\ {\scriptstyle FS}\downarrow & & \downarrow{\scriptstyle GS} \\ F\overline{Y} & \xrightarrow[\;t_{\overline{Y}}\;]{} & G\overline{Y}. \end{array}$$

This naturality, of course, implies that $t_{\overline{X}} \in L_{L\overline{X}}(F\overline{X},G\overline{X})$; i.e. t is a collection of module maps $\{t_{\overline{X}} \in L_{L\overline{X}}(F\overline{X},G\overline{X})\}$ with an extended naturality condition. In fact $NAT(F,G)$ is actually the projective limit of $L_{L\overline{X}}(F\overline{X},G\overline{X})$, $\overline{X} \in \overline{\mathfrak{B}}$; a point of view similar to this is followed in [16]. Given two natural transformations $s,t:F \rightrightarrows G$, we can define on $t_{\overline{X}}$ and $s_{\overline{X}}$ the obvious pointwise sum and scalar product. Thus, it is easy to see that the set of natural transformations $\mathfrak{B}^{\overline{\mathfrak{B}}}(F,G)$ forms the unit ball of a Banach space $NAT(F,G)$. The norm in $NAT(F,G)$ is given by $t = \sup\{\|t_{\overline{X}}\|:\overline{X} \in \overline{\mathfrak{B}}\}$, which can be shown to be finite.

In many important examples the set $NAT(F,G)$ is actually determined by a single diagram $\overline{X}$. This will be seen to be the case in Chapter VI when F is the left Kan extension Lan_A (or where G is the right Kan extension Ran_A). It is also the case for $F=\Delta$ and $G=\Sigma$, for we can see as follows that $NAT(\Delta,\Sigma) = L_{L\overline{I}}(\Delta\overline{I},\Sigma\overline{I}) = I$. If $t \in I$ is given with $|t| \leq 1$, we may define $t_{\overline{X}} \in L_{L\overline{X}}(\Delta\overline{X},\Sigma\overline{X})$ in precisely one natural way from t: namely if $x = (x_0,x_1) \in \Delta\overline{X}$, and $\hat{x}$ denotes the map from $\overline{I}$ to $\overline{X}$ sending $(1,1)$ to (x_0,x_1), then we must define $t_{\overline{X}}(x)$ in such a way that the following diagram is commutative:

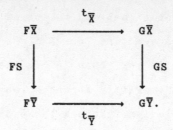

$$\begin{array}{ccc} I=\Delta\overline{I} & \xrightarrow{\quad t \quad} & \Sigma\overline{I}=I \\ {\scriptstyle \Delta\hat{x}}\Big\downarrow & & \Big\downarrow{\scriptstyle \Sigma\hat{x}} \\ \Delta\overline{X} & \xrightarrow[\quad t_{\overline{X}} \quad]{} & \Sigma\overline{X} \end{array},$$

so $t_{\overline{X}}(x) = \Sigma\hat{x}t(1) = tj(x)$. Recalling that Δ is $L(\overline{I},-)$, we can see that this example is a special case of the following proposition which is the enriched version (in the sense of $\mathscr{C}$-categories) of the well known Yoneda lemma.

4.4. <u>Proposition</u>. (Yoneda Lemma). Let $F:\overline{\mathscr{B}} \to \mathscr{B}$ be a functor and $\overline{X}\in\overline{\mathscr{B}}$. Then

$$NAT(L(\overline{X},-),F) = F\overline{X}.$$

<u>Proof</u>: Let $t \in NAT(L(\overline{X},-),F)$. Then $t_{\overline{X}} \in L_{L\overline{X}}(L(\overline{X},\overline{X}),F\overline{X})$, so $t_{\overline{X}}(1_{\overline{X}}) \in F\overline{X}$. Now we show that t is actually determined by this element $x = t_{\overline{X}}(1_{\overline{X}})$ due to its naturality. For any $\overline{Y}\in\overline{\mathscr{B}}$ and $S:\overline{X} \to \overline{Y}$, the following diagram must be commutative:

$$\begin{array}{ccc} L(\overline{X},\overline{X}) & \xrightarrow{\quad t_{\overline{X}} \quad} & F\overline{X} \\ {\scriptstyle L(X,S)}\Big\downarrow & & \Big\downarrow{\scriptstyle FS} \\ L(\overline{X},\overline{Y}) & \xrightarrow[\quad t_{\overline{Y}} \quad]{} & F\overline{Y} \end{array}.$$

This says that $t_{\overline{Y}}(S) = FS(x)$. □

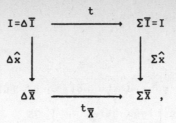

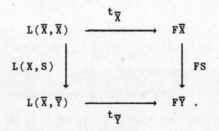

5. Interpolation Spaces and Functors.

In Chapter I we introduced the notion of interpolation spaces for $\overline{X}$. We repeat those definitions here.

I.3.1. Definitions. An $L(\overline{X})$-module X is called (i) a *quasi-interpolation space* for $\overline{X}$ if there are $L(\overline{X})$-module maps $\delta:\Delta\overline{X} \to X$ and $\sigma:X \to \Sigma\overline{X}$ such that $\sigma\circ\delta = j$; (ii) a *Δ-interpolation space* if it is a quasi-interpolation space with δ epi; (iii) a *Σ-interpolation space* if it is a quasi-interpolation space with σ mono.

5.1. Examples. (1) Clearly, $\Delta\overline{X}$ and $\Sigma\overline{X}$ are Δ- and Σ-interpolation spaces for $\overline{X}$, respectively. (2) While the only Δ- or Σ-interpolation space for the constant diagram JX is X, quasi-interpolation spaces would include the following type of example

$$\begin{array}{ccc} X & \longrightarrow & X \\ \Big\downarrow & \delta & \Big\downarrow \\ & X\Pi X & \\ & & \sigma \\ X & \longrightarrow & X , \end{array}$$

where δ is the diagonal map and σ is half the sum map. Thus, the Δ- or Σ-condition is necessary to eliminate this unnatural situation for interpolation theory. (3) We refer the reader to I.3.2 for an example of a Δ-interpolation space different from $\Delta\overline{X}$ which is not a Σ-interpolation space.

We would like to remark here that there is an alternative definition for Δ- and Σ-interpolation spaces in the case that $\overline{X}$ is unital as defined in Section 2 of this chapter. The import of the proposition is that in some sense the Δ-norm is the largest and the

Σ-norm the smallest with which an $L\overline{X}$-module may be endowed.

Before stating the next proposition it will be useful to introduce an additional piece of notation.

5.2. Notation. Let $\overline{X}$ be unital with unit u and dual unit u'. If $x \in \Delta\overline{X}$, then clearly $u' \otimes x$ defines an operator of norm less than or equal to $\|x\|$ on $\overline{X}$ such that

$$(u' \otimes x)(y) = \langle u', jy \rangle x$$

for $y \in \Delta\overline{X}$. Similarly, for any $x' \in (\Sigma\overline{X})'$, $x' \otimes u$ defines an operator on $\overline{X}$ of norm less than or equal to $\|x'\|$ such that $(x' \otimes u)(y) = \langle x', jy \rangle u$. We shall denote these operators as follows: $u' \otimes x = \hat{x}$ and $x' \otimes u = \hat{x}'$.

5.3. Proposition. Let $\overline{X}$ be unital with unit u and let X be an $L\overline{X}$-module. Then (1) X is a Δ-interpolation space for $\overline{X}$ if there is a module map $\delta : \Delta\overline{X} \to X$ which is epi and such that $\|\delta u\| = 1$; (2) X is a Σ-interpolation space for $\overline{X}$ if there is a module map $\sigma : X \to \Sigma\overline{X}$ which is mono and such that $\|\sigma\| = 1$.

Proof: (1) If we can show that for all $x \in \Delta\overline{X}$, $\|jx\|_{\Sigma\overline{X}} \leq \|\delta x\|_X$, then we can define $\sigma\delta x$ to be jx and extend σ to all of X since $\Delta\overline{X}$ is dense in X. To this end, let $x' \in (\Sigma\overline{X})'$ be such that $\|x'\| = 1$ and $\langle x', jx \rangle = \|jx\|$. Then $\hat{x}'(\delta x) = \langle x', jx \rangle \delta u = \|jx\| \delta u$. Hence, $\|jx\| = \|jx\| \, \|\delta u\| = \|\hat{x}'(\delta x)\| \leq \|\delta x\|$. (2) Let $x_0 \in X$ be nonzero. Choose $x_0' \in (\Sigma\overline{X})'$ such that $\langle x_0', \sigma x_0 \rangle = 1$ and define $\tilde{u} = \hat{x}_0'(x_0)$. We have $\tilde{u} = \hat{u}'(\tilde{u})$ since

$$\sigma\hat{u}'(\hat{x}_0'(x_0)) = \sigma\hat{u}'(\langle x_0', \sigma x_0 \rangle \delta u) = ju = \sigma\tilde{u},$$

so $\tilde{u}$ is a suitable substitute for u in X. We now define for any $x \in \Delta\overline{X}$, $\delta x = \hat{x}(\tilde{u})$. We may easily check that δ is a norm-decreasing module map and that $\sigma \circ \delta = j$. □

We remark that the above proposition is topologically true for non-trivial $\overline{X}$ (i.e. $j \neq 0$) without the requirements placed on $\|\delta u\|$ and $\|\sigma\|$.

Turning now to functors $F:\overline{\mathcal{B}} \to \mathcal{B}$, we notice that $F\overline{X}$ is as always an $L\overline{X}$-module by definition of functor. Furthermore, if the condition $F \circ J = 1_{\mathcal{B}}$ is met, i.e. if F is constant on constant diagrams, then maps $\delta:\Delta\overline{X} \to F\overline{X}$ and $\sigma:F\overline{X} \to \Sigma\overline{X}$ arise as follows which automatically make $F\overline{X}$ into a quasi-interpolation space: we have natural transformations

$$J\Delta\overline{X} \xrightarrow{(\delta_0,\delta_1)} \overline{X} \xrightarrow{(\sigma_0,\sigma_1)} J\Sigma\overline{X}$$

and, hence, module maps

$$FJ\Delta\overline{X} = \Delta\overline{X} \xrightarrow{F(\delta_0,\delta_1)=\delta} F\overline{X} \xrightarrow{F(\sigma_0,\sigma_1)=\sigma} FJ\Sigma\overline{X} = \Sigma\overline{X}.$$

From this discussion the following definitions appear very natural.

5.4. Definitions. A functor $F:\overline{\mathcal{B}} \to \mathcal{B}$ is said to be (i) a quasi-interpolation functor if $F \circ J = 1_{\mathcal{B}}$; a Δ-interpolation functor if F is a quasi-interpolation functor such that $F\overline{X}$ is a Δ-interpolation space for each $\overline{X}$; (iii) a Σ-interpolation functor if F is a quasi-interpolation functor such that $F\overline{X}$ is a Σ-interpolation space for all $\overline{X}$.

There is a strong connection between our definitions 4.2 and the above definitions.

5.5. Proposition. Let $F:\overline{\mathfrak{B}} \to \mathfrak{B}$ be a quasi-interpolation functor. 1. F is a Δ-interpolation functor if and only if F is a Δ-functor. 2. F is a Σ-interpolation functor if and only if F is a Σ-functor such that the coevaluation map

$$\eta:F\overline{X} \to L(L(\overline{X},\overline{I}),F\overline{I}) = \Sigma\overline{X}''$$

factors through $\Sigma\overline{X}$.

Proof: The first equivalence follows directly from the fact that when F is a quasi-interpolation functor

$$\varepsilon:L(\overline{I},\overline{X})\otimes F\overline{I} = \Delta\overline{X} \to F\overline{X}$$

is simply δ. The second also follows directly from the definitions.□

It is redundant to note that Δ and Σ are the paradigms of Δ- and Σ-functors.

Given a quasi-interpolation functor F, there is always a natural way to construct Δ- and Σ-interpolation functors from F, which connects our theory with the classical theory of interpolation. We define

$$F^{o}\overline{X} = \mathscr{C}\ell(\mathrm{im}\ \delta)$$

and

$$F^{s}\overline{X} = F\overline{X}/\ker\ \sigma.$$

Then the following proposition is completely obvious.

5.6. Proposition. Let F be a quasi-interpolation functor. Then F^o and F^s are Δ-interpolation and Σ-interpolation functors, respectively. Moreover, if F is a Δ-interpolation (Σ-interpolation) functor, then F^s (F^o) is both Δ- and Σ-interpolation.

Finally, because of the difficulty that the dual space of an interpolation space X for $\overline{X}$ need not be an interpolation space for $\overline{X}'$, we wish to close this chapter by introducing the notion of a canonical dual for interpolation spaces. We recall that even to have X' be an intermediate space for $\overline{X}'$ in the classical theory required that $\Delta\overline{X}$ be dense in X (i.e. in our terminology that X be a Δ-interpolation space). However, even in this case it does not follow that X' is an $L\overline{X}'$-module. Thus, we wish to define a reasonable substitute for X' which will be an $L\overline{X}'$-module. It is natural that Banach module homomorphisms enter into this definition.

5.7. Definition. Let X be an $L\overline{X}$-module. The interpolation dual X^* of X is defined by

$$X^* = L_{L\overline{X}}(X, \overline{X}\otimes\overline{X}').$$

We see that if $f: X \to Y$ is an $L\overline{X}$-module homomorphism, then there is a dual map $f^*: Y^* \to X^*$, so the operation $(\ )^*$ is functorial on the level of $L\overline{X}$-modules.

We observe first that the interpolation dual has the following properties.

5.8. Proposition. Let X be an $L\overline{X}$-module. Then X^* is an $L\overline{X}'$-module. Moreover, we have

$$(\Delta\overline{X})^* = \Sigma\overline{X}'.$$

<u>Proof</u>: That X^* is an $L\overline{X}'$-module follows easily from the bimodule structure of $\overline{X}\otimes\overline{X}'$: we define for $T:\overline{X}' \to \overline{X}'$ and $f \in L_{L\overline{X}}(X, \overline{X}\otimes\overline{X}')$, $Tf = (\overline{X}\otimes T)\circ f$. Now assume for convenience that $\overline{X}$ is unital. Let $\alpha \in (\Delta\overline{X})^* = L_{L\overline{X}}(\Delta\overline{X}, \overline{X}\otimes\overline{X}')$. Then we can choose a representation for $\alpha(u) \in \overline{X}\otimes\overline{X}'$ as

$$\Sigma x_{0i}\otimes x'_{0i} + \Sigma x_{1i}\otimes x'_{1i},$$

where $x_{ji} \in X_j$ and $x'_{ji} \in X'_j$ for $j=0,1$. Now for $x \in \Delta\overline{X}$. we let $\hat{x}$ denote the operator on $\overline{X}$ defined in 5.2. Then since α is an $L\overline{X}$-module map, we have

$$\begin{aligned}\alpha(x) &= \alpha(\hat{x}(u)) = \hat{x}\otimes\overline{X}'(\alpha(u)) \\ &= \hat{x}\otimes\overline{X}'(\Sigma x_{0i}\otimes x'_{0i} + \Sigma x_{1i}\otimes x'_{1i}) \\ &= \Sigma\langle u', \sigma_{0i}x_{0i}\rangle\delta_0 x\otimes x'_{0i} + \Sigma\langle u', \sigma_{1i}x_{1i}\rangle\delta_1 x\otimes x'_{1i} \\ &= \Sigma\delta_0 x\otimes\langle u', \sigma_{0i}x_{0i}\rangle x'_{0i} + \Sigma\delta_1 x\otimes\langle u', \sigma_{1i}x_{1i}\rangle x'_{1i} \\ &= \delta_0 x\otimes x'_0 + \delta_1 x\otimes x'_1,\end{aligned}$$

where

$$x'_j = \Sigma\langle u', \sigma_{ji}x_{ji}\rangle x'_{ji}, \quad j=0,1.$$

Hence, α depends only on $\sigma_0 x'_0 + \sigma_1 x'_1 \in \Sigma\overline{X}'$. Conversely, if $\beta \in \Sigma\overline{X}'$, then we can write $\beta = x'_0 + x'_1$ for $x'_j \in X'_j$, $j=0,1$, and we can define $\overline{\beta}(x) = x\otimes\beta = \delta_0 x\otimes x'_0 + \delta_1 x\otimes x'_1$ for $x \in \Delta\overline{X}$. This shows that $(\Delta\overline{X})^* = \Sigma\overline{X}'$. With some modifications in the proof for non-unital $\overline{X}$, the proof works in general. □

Furthermore, we have the following characterization of X^* when X is a Δ-interpolation space for $\overline{X}$.

5.9. <u>Proposition</u>. If X is a Δ-interpolation space for the doolittle diagram $\overline{X}$, then

$$X^* \subset X' \subset \Sigma\overline{X}'$$

and X^* is the maximal $L\overline{X}'$-module contained in X'.

<u>Proof</u>: Since $\delta:\Delta\overline{X} \to X$ is epi, we immediately have $X' \subset (\Delta\overline{X})' = \Sigma\overline{X}'$. Moreover, it is obvious that the map

$$\delta^* = L_{L\overline{X}}(\delta, \overline{X}\otimes\overline{X}') : X^* = L_{L\overline{X}}(X, \overline{X}\otimes\overline{X}') \to L_{L\overline{X}}(\Delta\overline{X}, \overline{X}\otimes\overline{X}') = \Sigma\overline{X}'$$

is a mono. However, it is clear from an examination of the identification in 5.8 of $(\Delta\overline{X})^*$ and $\Sigma\overline{X}'$ that $\delta^* = \delta'\circ\tau$, where $\tau : X^* \to X'$ is given by $\tau(\alpha) = \mathrm{Tr}\circ\alpha$. Therefore, τ is a mono, so $X^* \subset X'$.

To see that X^* is the maximal $L\overline{X}'$-module contained in X', let $x' \in X' \subset \Sigma\overline{X}'$. Then $x' \in X^*$ if and only if for all $x \in \Delta\overline{X}$

$$\|x\otimes x'\|_{\overline{X}\otimes\overline{X}'} \le C\|x\|_X.$$

However,

$$\begin{aligned} \|x\otimes x'\|_{\overline{X}\otimes\overline{X}'} &= \sup\{|\langle T, x\otimes x'\rangle| \,\big|\, T \in L(\overline{X}'),\ \|T\| \le 1\} \\ &= \sup\{|\langle Tx', x\rangle| \mid T \in L(\overline{X}'),\ \|T\| \le 1\}. \end{aligned}$$

Therefore, if $\sup\{\|Tx'\|_{\overline{X}'} \mid T \in L(\overline{X}'),\ \|T\| \le 1\} < \infty$, then $x' \in X^*$ and $\|x'\|_{X^*} = \sup\{\|Tx'\|\}$. So X^* consists of those $x' \in X'$ for which

$Tx' \in X'$ for all $T \in L(\overline{X}')$, and this is clearly the maximal $L(\overline{X}')$-module contained in X'. □

5.10. <u>Corollary</u>. If $\Delta\overline{X}$ is dense in $\Sigma\overline{X}$, then $(\Sigma\overline{X})^* = \Delta(\overline{X}')$.

Finally, we can prove that the previous corollary is true even without the density condition.

5.11. <u>Proposition</u>. Let $\overline{X} \in \overline{\mathfrak{B}}$. Then (i) $X_i^* = X_i'$, $i=0,1$, and (ii) $(\Sigma\overline{X})^* = \Delta(\overline{X}')$.

<u>Proof</u>: (i) Let us take $i=0$. We assume that $\Delta\overline{X}$ is not dense in X_0, since otherwise the result holds by 5.9. Then we can find $u_0 \in X_0$ such that $\|u_0\| < 1 + \varepsilon$ and $\inf\{\|u_0 - \delta_0 x\| \mid x \in \Delta\overline{X}\} \geq 1$. We choose $f \in X_0'$ such that $\langle f, u_0\rangle = 1$ and $f \in \operatorname{im}(\delta_0)^{\perp}$. We note that for each $x_0 \in X_0$, the map $\hat{x}_0 : X_0 \to X_0$ defined by $\hat{x}_0(y) = \langle f, y\rangle x_0$ gives a map $(\hat{x}_0, 0) : \overline{X} \to \overline{X}$ and that $\hat{x}_0(u_0) = x_0$. Now if $\alpha \in X_0^* = L_{L\overline{X}}(X_0, \overline{X}\otimes\overline{X}')$, we let $\alpha(u_0)$ be represented by

$$\Sigma x_{0i} \otimes x_{0i}' + \Sigma x_{1i} \otimes x_{1i}'.$$

Then

$$\begin{aligned}
\alpha(x_0) &= \alpha(\hat{x}_0(u_0)) = (\hat{x}_0 \otimes \overline{X}')(\alpha(u_0)) \\
&= (\hat{x}_0 \otimes \overline{X}')(\Sigma x_{0i} \otimes x_{0i}' + \Sigma x_{1i} \otimes x_{1i}') \\
&= \Sigma \langle f, x_{0i}\rangle x_0 \otimes x_{0i}' + 0 \\
&= x_0 \otimes \Sigma \langle f, x_{0i}\rangle x_{0i}' = x_0 \otimes x_0',
\end{aligned}$$

where $x_0' = \Sigma\langle f, x_{0i}\rangle x_{0i}'$. Therefore, α is determined by the element x_0' in X_0'. Evidently, any element in X_0' similarly determines an element of X_0^*, so this proves (i).

The result (ii) follows from (i) since $L_{L\overline{X}}(\Sigma\overline{X}, \overline{X}\otimes\overline{X}') = (\Sigma\overline{X})^*$

must be the pullback of the diagram

$$\begin{array}{ccc} & & X_0^* = X_0' \\ & & \big\downarrow \\ X_1' = X_1^* & \longrightarrow & (\Delta\overline{X})^* = \Sigma\overline{X}' \end{array},$$

which is $\Delta\overline{X}'$. □

CHAPTER V

FINITE DIMENSIONAL DOOLITTLE DIAGRAMS

1. <u>1-dimensional Doolittle Diagrams and Applications</u>.

The simplest objects in any category often play a special role. In the category $\overline{\mathfrak{D}}$ the simplest objects are the 1-dimensional objects, i.e. diagrams $\overline{X}$ such that all the spaces $\Delta\overline{X}$, X_0, X_1, $\Sigma\overline{X}$ are at most 1-dimensional. These diagrams do play a special role, and we shall therefore give them individual names.

1.1. <u>Definitions</u>. 1. Let $\overline{K}_0$ denote the diagram

$$\overline{K}_0 = \begin{array}{ccc} I & \xrightarrow{\;1\;} & I \\ \downarrow & & \downarrow \\ 0 & \longrightarrow & 0 \end{array}$$

and let $\overline{K}_1 = (\overline{K}_0)^\tau$, the transposed diagram (IV.1.3). 2. Let $\overline{P}_0$ denote the diagram

$$\overline{P}_0 = \begin{array}{ccc} 0 & \longrightarrow & I \\ \downarrow & & \downarrow{\scriptstyle 1} \\ 0 & \longrightarrow & I \end{array}$$

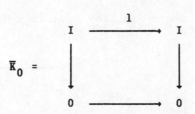

and let $\overline{P}_1 = (\overline{P}_0)^{\tau}$. 3. Let $\overline{I}(s,t)$ denote the diagram

$$\begin{array}{ccc} I & \xrightarrow{\;s\;} & I \\ {\scriptstyle t}\big\downarrow & & \big\downarrow{\scriptstyle t} \\ I & \xrightarrow[\;s\;]{} & I \end{array} \quad ,$$

where s and t are real numbers such that $|s|, |t| \leq 1$ and $\max(s,t) = 1$. We denote $\overline{I}(-1,1)$ by $\overline{I}_0^-$, $\overline{I}(1,-1)$ by $\overline{I}_1^-$, and $\overline{I}(1,1)$, as before, simply by $\overline{I}$.

1.2. <u>Proposition</u>. The above examples are (up to isomorphism) the only 1-dimensional doolittle diagrams in $\mathfrak{B}$.

<u>Proof</u>: It would be conceivable that $\overline{I}(s,t)$ could be defined even if $\max(s,t) < 1$, but in this case it is easily seen that $\overline{I}(s,t)$ is neither a pushout nor a pullback diagram. □

Besides the 1-dimensional diagrams also the following diagrams $\overline{K} = \overline{K}_0 \sqcap \overline{K}_1$ and $\overline{P} = \overline{P}_0 \sqcup \overline{P}_1$ are useful for certain constructions:

$$\overline{K} = \begin{array}{ccc} I^2 & \longrightarrow & I \\ \big\downarrow & & \big\downarrow \\ I & \longrightarrow & 0 \end{array} \quad , \qquad \overline{P} = \begin{array}{ccc} 0 & \longrightarrow & I \\ \big\downarrow & & \big\downarrow \\ I & \longrightarrow & I^2 \end{array} \quad .$$

We remark that $\overline{K}_i' = \overline{P}_i$, $\overline{P}_i' = K_i$, $\overline{K}' = \overline{P}$, $\overline{P}' = \overline{K}$, and $\overline{I}(s,t)' = \overline{I}(t,s)$. We shall next define certain natural and useful functors that arise from the above diagrams.

1.3. Definitions. Let $\overline{X}\in\overline{\mathfrak{B}}$. We shall denote:

(i) $K_i\overline{X} = L(\overline{K}_i,\overline{X})$;

(ii) $K\overline{X} = L(\overline{K},\overline{X})$;

(iii) $X_i/\Delta = \overline{K}_i\otimes\overline{X}$;

(iv) $\Sigma\overline{X}/\Delta = \overline{K}\otimes\overline{X}$;

(v) $P_i\overline{X} = L(\overline{P}_i,\overline{X}) = \overline{P}_i\otimes\overline{X}$;

(vi) $P_\pi\overline{X} = L(\overline{P},\overline{X})$;

(vii) $P_\amalg\overline{X} = \overline{P}\otimes\overline{X}$;

(viii) $\Delta(s,t,\overline{X}) = L(\overline{I}(s,t),\overline{X})$, $s \neq -1$;

(ix) $\Delta_0^-\overline{X} = L(\overline{I}_0^-,\overline{X})$;

(x) $\Sigma(s,t,\overline{X}) = \overline{I}(s,t)\otimes\overline{X}$.

These functors can easily be analyzed as follows.

1.4. Proposition. Let $\overline{X}\in\overline{\mathfrak{B}}$. Then

(1) $K_i\overline{X}$ is isomorphic to $\ker(\sigma_i) \subset X_i$;

(2) $K\overline{X} = \ker(j)$;

(3) X_i/Δ is the quotient space of X_i with respect to the (norm) closure of $\mathrm{im}(\delta_i)$ in X_i;

(4) $\Sigma\overline{X}/\Delta$ is the quotient space of $\Sigma\overline{X}$ with respect to the (norm) closure of $\mathrm{im}(j)$ in $\Sigma\overline{X}$;

(5) $P_i\overline{X} = X_i$;

(6) $P_\pi\overline{X} = X_0\pi X_1$, the Banach product space (see I.1);

(7) $P_\amalg\overline{X} = X_0\amalg X_1$, the Banach sum space (see I.1).

Proof: Since the proofs of the above statements are similar, we shall verify 1 and 3 and leave the remaining verifications to the reader.

By definition $K_0\overline{X} = L(\overline{K}_0,\overline{X})$ is the pullback of the diagram

$$\begin{array}{ccc} & & L(I,X_0) \\ & & \downarrow \\ L(0,X_1) & \longrightarrow & L(I,\Sigma\overline{X}) \end{array},$$

i.e. of the diagram

$$\begin{array}{ccc} & & X_0 \\ & & \big\downarrow{\scriptstyle \sigma_0} \\ 0 & \longrightarrow & \Sigma\overline{X}\ , \end{array}$$

and this is obviously $\ker(\sigma_0)$, which proves (1).

On the other hand, $X_0/\Delta = \overline{K}_0\otimes\overline{X}$ is the pushout of the diagram

$$\begin{array}{ccc} I\otimes\Delta\overline{X} & \longrightarrow & I\otimes X_0 \\ \big\downarrow & & \\ 0\otimes X_1 & & , \end{array}$$

i.e. of the diagram

$$\begin{array}{ccc} \Delta\overline{X} & \xrightarrow{\ \delta_0\ } & X_0 \\ \big\downarrow & & \\ 0 & & , \end{array}$$

and this is $X_0/\mathcal{C}\ell(\mathrm{im}\,\delta_0)$. □

1.5. <u>Remark</u>. Statements 1 and 2 of the above proposition show that there is agreement with the definitions of $K_i\overline{X}$ and $K\overline{X}$ given in I.2.

There is a connection between the functors $\Delta(s,t,-)$ and $\Sigma(s,t,-)$ and the K- and J-functionals of classical interpolation theory. To see this we first recall the definitions of Lions-Peetre.

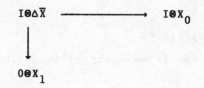

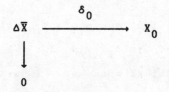

II.1.1. <u>Definition</u>. Let $\overline{X}$ be a Banach couple (i.e. a doolittle

diagram with all maps monomorphisms). Let t be a positive real number. Then we define

$$J(t,x) = \sup\{\|x_0\|_{X_0},\ t\|x_1\|_{X_1} \mid (x_0,x_1) = x\in\Delta\overline{X}\}$$

and

$$K(t,x) = \inf\{\|x_0\|_{X_0} + t\|x_1\|_{X_1} \mid x\in\Sigma\overline{X},\ x = x_0+x_1\}.$$

We now have the following proposition which indicates that $\Delta(s,t,\overline{X})$ and $\Sigma(s,t,\overline{X})$ may be considered as "generalizations of the J- and K-functionals".

1.6. Proposition. Let $\overline{X}$ be a Banach couple and let $0<s,t\leq 1$ with $\max(s,t) = 1$. Then (i) $\Delta(s,t,\overline{X})$ is isomorphic to the linear space $\Delta\overline{X}$ considered as a Banach space with the norm

$$\|x\|_{\Delta(s,t,\overline{X})} = 1/s\ J(s/t,\ x),$$

and (ii) $\Sigma(s,t,\overline{X})$ is isomorphic to the linear space $\Sigma\overline{X}$ considered as a Banach space with the norm

$$\|x\|_{\Sigma(s,t,\overline{X})} = sK(t/s,\ x).$$

Proof: (1) We have $\Delta(s,t,\overline{X}) = L(\overline{I}(s,t),\ \overline{X})$, so let $T = (T_0,T_1):\overline{I}(s,t) \to \overline{X}$. Then $T_i:I \to X_i$, $i=0,1$, so denoting $T_i(1)$ by $x_i\in X_i$, we have $sx_0 = tx_1 = x\in\Sigma\overline{X}$. Now obviously $x \in X_0 \cap X_1 = \Delta\overline{X}$ and

$$\begin{aligned} \|T\| &= \max(\|x_0\|_{X_0}, \|x_1\|_{X_1}) \\ &= \max(1/s\|x\|_{X_0}, 1/t\|x\|_{X_1}) \\ &= 1/s\ J(s/t, x). \end{aligned}$$

(ii) We have $\Sigma(s,t,\overline{X}) = \overline{I}(s,t)\otimes\overline{X}$, which is by definition the pushout of the diagram

$$\begin{array}{ccc} I\otimes\Delta\overline{X} & \xrightarrow{\ s\delta_0\ } & I\otimes X_0 \\ {\scriptstyle t\delta_1}\big\downarrow & & \\ I\otimes X_1 & & . \end{array}$$

Therefore, we have $\Sigma(s,t,\overline{X}) = (X_0 \amalg X_1)/N$, where $N = \{(x_0,x_1) | \exists d \in \Delta,\ x_0 = sd,\ x_1 = -td\}$. This means that if $y \in \Sigma(s,t,\overline{X})$, then there exist $x_0 \in X_0$, $x_1 \in X_1$ such that $y = (x_0,x_1) + N$, and

$$\begin{aligned} \|y\|_{\Sigma(s,t,\overline{X})} &= \inf\{\|x_0+sd\|_{X_0} + \|x_1-td\|_{X_1} | d \in \Delta\overline{X}\} \\ &= \inf\{s\|x_0/s + d\|_{X_0} + t\|x_1/t - d\|_{X_1} | d \in \Delta\overline{X}\}. \end{aligned}$$

Therefore, we may define

$$y = x_0/s + x_1/t \in \Sigma\overline{X},$$

and we see that

$$\begin{aligned} \|y\|_{\Sigma(s,t,\overline{X})} &= \inf\{s\|y_0\|_{X_0} + t\|y_1\|_{X_1} | y_0 + y_1 = y\} \\ &= s\ \inf\{\|y_0\|_{X_0} + t/s\|y_1\|_{X_1} | y_0 + y_1 = y\} \\ &= s\ K(t/s, y). \end{aligned}$$

□

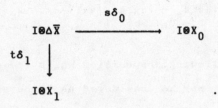

It should be observed that for general doolittle diagrams, $\Delta(s,t,\overline{X})$ and $\Sigma(s,t,\overline{X})$ are $\mathfrak{B}_\infty$-isomorphic to $\Delta\overline{X}$ and $\Sigma\overline{X}$, respectively.

1.7. Remark. It follows, by IV.2.3, from the adjointness of $\overline{L}(\overline{X},J(-))$ and $\overline{X}\otimes-$, that $L(\overline{A},\overline{X}')$ is the dual space $(\overline{A}\otimes\overline{X})'$. However, we notice in the description of the above functors that $\overline{P}_i\otimes\overline{X}'$, $\overline{P}\otimes\overline{X}'$, $\Delta(s,t,\overline{X}')$ and $\Sigma(s,t,\overline{X}')$ are also dual spaces. Only $\overline{K}_i\otimes\overline{X}'$ and $\overline{K}\otimes\overline{X}$ fail to be dual spaces. We shall presently prove in Section 4 more generally that $\overline{E}\otimes\overline{X}' = L(\overline{E},\overline{X})'$ when $\overline{E}$ is a finite dimensional Banach couple.

2. The Structure Theorem.

Finite dimensional Banach spaces are important not only in finite-dimensional problems but also in the general theory of Banach spaces. One reason for this is the simple fact that every Banach space is the inductive limit of its finite-dimensional subspaces. For the same reason finite-dimensional doolittle diagrams are useful in general interpolation theory - in particular in connection with duality problems.

The most useful technical device for studying problems concerning finite dimensional vector spaces (with or without additional structure) is the choice of a suitable basis. An example of a particularly useful basis in the study of finite dimensional Banach spaces is given by the following proposition, a proof of which may be found in [24].

2.1. Proposition. (Auerbach) Let X be an n-dimensional Banach space. Then there exists a basis $\{e_i\}_{i=1}^n$ with dual basis $\{e_i'\}_{i=1}^n$ such that $\|e_i\| = \|e_i'\| = 1$.

We remark that such a basis is called a <u>normal basis</u>. Analytically, for an n-dimensional space X it means that there exist norm-decreasing linear maps $\ell_n^1 \xrightarrow{U} X \xrightarrow{V} \ell_n^\infty$ such that $V\circ U$ is the canonical inclusion.

In the setting of doolittle diagrams we also want to have for each finite dimensional diagram a notion of basis which is faithful to the structure of the diagram and admits a dual basis of a similar type. We shall try to obtain an analogy to the normal basis of Auerbach. The effort to construct a normal basis for doolittle diagrams will reveal some of the intrinsic metric properties of this category.

First we will find it useful to have names for the subcategories of finite dimensional doolittle diagrams and couples.

2.2. <u>Definitions</u>. 1. Let $\overline{\mathcal{F}}$ denote the full subcategory of $\overline{\mathcal{B}}$ consisting of finite dimensional doolittle diagrams, i.e. diagrams $\overline{F}$ such that $\Delta\overline{F}$, F_0, F_1 and $\Sigma\overline{F}$ are finite dimensional. 2. Let $\overline{\mathcal{E}}$ denote the full subcategory of finite dimensional Banach couples.

Given an object $\overline{F}\in\overline{\mathcal{F}}$ we shall now try to define a basis which is as close as possible to being a normal basis for each of the spaces $\Delta\overline{F}$, F_0, F_1, and $\Sigma\overline{F}$. Let us write

$$\begin{aligned} k_i &= \dim(K_i\overline{F}), \\ d &= \dim(\Delta\overline{F}/K\overline{F}), \text{ and} \\ p_i &= \dim(F_i/\mathrm{im}(\delta_i)), \end{aligned}$$

so that we have

$$\dim \Delta\overline{F} = k_0 + k_1 + d,$$
$$\dim F_i = k_i + d + p_i, \text{ and}$$
$$\dim \Sigma\overline{F} = p_0 + p_1 + d.$$

Consequently, since $(F_i/\mathrm{im}(\delta_i))' = K_i\overline{F}'$ and $(K_i\overline{F})' = F_i'/\mathrm{im}(\sigma_i')$, we have $\dim(K_i\overline{F}') = p_i$ and $\dim(F_i'/\mathrm{im}\,\sigma_i') = k_i$. To construct compatible bases for the various spaces, we shall start by choosing vectors $\{k_{ij}'\}_{j=1}^{k_i}$ in F_i' such that $\|k_{ij}'\| = 1$ and $\{[k_{ij}' + \mathrm{im}\,\sigma_i']\}$ is a normal basis for $(K_i\overline{F})'$. Then the dual basis $\{k_{ij}\}_{j=1}^{k_i}$ is a normal basis for $K_i\overline{F}$. We shall denote by $\mathrm{Ann}(K')$ the set of all $x \in \Delta\overline{F}$ such that for all i and j,

$$\langle k_{ij}', \delta_i x\rangle = 0.$$

We can then choose a normal basis $\{d_j\}_{j=1}^{d}$ for $\mathrm{Ann}(K')$. In this way we obtain a basis

$$\{\{k_{0j}\}_{j=1}^{k_0} \cup \{k_{1j}\}_{j=1}^{k_1} \cup \{d_j\}_{j=1}^{d}\}$$

for $\Delta\overline{F}$ such that $\{\{k_{ij}\}_{j=1}^{k_i}\}_{i=0}^{1}$, is a normal basis for $K\overline{F}$.

Next we denote by $\mathrm{Ann}(K_i')$ the set of all $x_i \in F_i$ such that $\langle k_{ij}', x_i\rangle = 0$ for all $j=1,\ldots,k_i$. Then we choose $\{p_{ij}\}_{j=1}^{p_i}$ in F_i such that $\|p_{ij}\| = 1$ and $\{[p_{ij} + \mathrm{im}(\delta_i)]\}$ is a normal basis for $\mathrm{Ann}(K_i')/\mathrm{im}(\delta_i)$. It is clear that the set

$$\{\{k_{ij}\}_{j=1}^{k_i} \cup \{\delta_i(d_j)\}_{j=1}^{d} \cup \{p_{ij}\}_{j=1}^{p_i}\}$$

forms a basis for the space F_i such that $\|k_{ij}\| = \|p_{ij}\| = 1$ and $\|\delta_i(d_j)\| \le 1$.

Let us consider the dual basis for F_i'. We observe first that for our original vectors $\{k_{ij}'\}_{j=1}^{k_i}$ we have orthogonality with $\{d_j\}_{j=1}^{d}$, and we have $\langle k_{ij}', k_{im}\rangle = \begin{cases}1 & j=m\\ 0 & j\neq m\end{cases}$. Hence, the vectors $\{k_{ij}'\}_{j=1}^{k_i}$ are part of the dual basis. Moreover, we may consider the dual vectors $\{p_{ij}'\}_{j=1}^{p_i}$. It follows that $\{p_{ij}'\}_{j=1}^{p_i}$ is orthogonal to all vectors $\{\delta_i(k_{ij})\}$ and $\{\delta_i(d_j)\}$, so that

$$\{p_{ij}'\}_{j=1}^{p_i} \subset \operatorname{im}(\delta_i)^{\perp} = K_i(\overline{F}').$$

Furthermore, if we let $d_{ij} = \delta_i(d_j)$, $i=0,1$, then $\langle d_{ij}', \delta(k_{im})\rangle = 0$ for $i=0,1$, $j=1,\ldots,d$, and $m=1,\ldots,k_i$, while

$$\langle d_{ij}', \delta_i(d_m)\rangle = \begin{cases}1 & j=m\\ 0 & j\neq m\end{cases}.$$

Hence, for any $x\in\Delta\overline{F}$,

$$\langle \delta_0'(d_{0j}'), x\rangle = \langle d_{0j}', \delta_0 x\rangle = \langle d_{1j}', \delta_1 x\rangle = \langle \delta_1'(d_{1j}'), x\rangle,$$

i.e. we have $\delta_0'(d_{0j}') = \delta_1'(d_{1j}')$, so that in fact there exists $d_j'\in\Delta(\overline{F}')$ such that $\sigma_0'(d_j') = d_{0j}'$ and $\sigma_1'(d_j') = d_{1j}'$. Collecting the above results, we have proved the following which yields as a corollary a structure theorem for $\overline{\mathcal{F}}$.

2.3. <u>Proposition</u>. Let $\overline{F}$ be a finite dimensional doolittle diagram. Then there exist vectors $\{\{k_{ij}\}_{j=1}^{k_i}\}_{i=1,0}$, $\{d_j\}_{j=1}^{d}$, $\{\{p_{ij}\}_{j=1}^{p_i}\}_{i=0,1}$ such that

(i) $\{k_{ij}\}_{j=1}^{k_i}$ is a normal basis for $K_i\overline{F}$,

(ii) $\{k_{0j}\}_{j=1}^{k_0} \cup \{k_{1j}\}_{j=1}^{k_1} \cup \{d_j\}_{j=1}^{d}$ is a basis for $\Delta\overline{F}$,

(iii) $\{\delta_i(k_{ij})\}_{j=1}^{k_i} \cup \{\delta_i(d_j)\}_{j=1}^{d} \cup \{p_{ij}\}_{j=1}^{p_i}$ is a basis for F_i, and

(iv) $\{\sigma_0(p_{0j})\}_{j=1}^{p_0} \cup \{\sigma_1(p_{1j})\}_{j=1}^{p_1} \cup \{j(d_m)\}_{m=1}^{d}$ is a basis for $\Sigma\overline{F}$.

Furthermore, the dual basis for F'_i can be written as

$$\{\sigma'_i(p'_{ij})\}_{j=1}^{p_i} \cup \{\sigma'_i(d'_j)\}_{j=1}^{d} \cup \{k'_{ij}\}_{j=1}^{k_i}$$

and as above,

(i) $\{p'_{ij}\}_{j=1}^{p_i}$ is a basis for $K_i\overline{F}'$,

(ii) $\{p'_{0j}\}_{j=1}^{p_0} \cup \{p'_{1j}\}_{j=1}^{p_1} \cup \{d'_j\}_{j=1}^{d}$ is a basis for $\Delta\overline{F}'$, and

(iii) $\{\delta'_0 k_{0j}\}_{j=1}^{k_0} \cup \{\delta'_1 k_{1j}\}_{j=1}^{k_1} \cup \{j'(d'_m)\}_{m=1}^{d}$ is a basis for $\Sigma\overline{F}'$.

2.4. Corollary. For any $\overline{F}\in\mathcal{F}$ there exist non-negative integers, d, k_0, k_1, p_0, and p_1 such that as doolittle diagrams of vector spaces, there is an isomorphism

$$\overline{F} \approx d\cdot\overline{I} \amalg k_0\cdot\overline{K}_0 \amalg k_1\cdot\overline{K}_1 \amalg p_0\cdot\overline{P}_0 \amalg p_1\cdot\overline{P}_1,$$

and the numbers d, k_i, and p_i satisfy the relations:

$$\dim \Delta\overline{F} = d + k_0 + k_1,$$
$$\dim F_i = k_i + d + p_i, \text{ and}$$
$$\dim \Sigma\overline{F} = d + p_0 + p_1.$$

3. Operators of Finite Rank

3.1. Definition. If $\overline{X}$ and $\overline{Y}$ are doolittle diagrams, we shall say that $T = (T_0, T_1) \in L(\overline{X}, \overline{Y})$ is of rank 1 if T factors through a 1-dimensional doolittle diagram.

In terms of this definition and our description of 1-dimensional doolittle diagrams (1.1 and 1.2) there are, thus, six cases of rank 1 operators, namely $T = V \circ U: \overline{X} \to \overline{F} \to \overline{Y}$ where $\overline{F}$ is $\overline{I}$, $\overline{I}(s,t)$, $\overline{K}_0$, $\overline{K}_1$, $\overline{P}_0$ or $\overline{P}_1$. If $\overline{F} = \overline{I}$, then $U \in L(\overline{X}, \overline{I}) = \Delta\overline{X}'$ and $V \in L(\overline{I}, \overline{Y}) = \Delta\overline{Y}$, so $T \sim x' \otimes y$, where $x' \in \Delta\overline{X}'$ and $y \in \Delta\overline{Y}$. Similarly, if $\overline{F} = \overline{I}(s,t)$ with $\min(s,t) \neq 0$, then

$$U \in L(\overline{X}, \overline{I}(s,t)) = L(\overline{I}(s,t)', \overline{X}') = L(\overline{I}(t,s), \overline{X}') = \Delta(t,s,\overline{X}')$$

and

$$V \in L(\overline{I}(s,t), \overline{Y}) = \Delta(s,t,\overline{Y}).$$

Moreover, in $\mathfrak{B}_\infty$, $\Delta(t,s,\overline{X}') = \Delta\overline{X}'$ and $\Delta(s,t,\overline{Y}) = \Delta\overline{Y}$, so qualitatively this case is the same as the previous case. If $\overline{F}$ is $\overline{K}_i$, $i=0$ or 1, then U may be identified with some element of X_i' since $L(\overline{X}, \overline{K}_i) = L(\overline{K}_i', \overline{X}') = L(\overline{P}_i, \overline{X}') = X_i'$ by 1.3. On the other hand, $V \in L(\overline{K}_i, \overline{Y}) = K_i\overline{Y}$, so $T \sim x_i' \otimes k_i$, where $x_i' \in X_i'$ and $k_i \in K_i\overline{Y}$. Finally, if $\overline{F} = \overline{P}_i$, $i=0$ or 1, then $L(\overline{X}, \overline{P}_i) = L(\overline{K}_i, \overline{X}') = K_i(\overline{X}')$, while $L(\overline{P}_i, \overline{Y}) = Y_i$. Hence, $T \sim k_i' \otimes y_i$, where $k_i' \in K_i(\overline{X}')$ and $y_i \in Y_i$.

Having now analyzed the operators of rank 1, we are in a better position to discuss operators of finite rank.

3.2. Definition. If $\overline{X}$ and $\overline{Y}$ are doolittle diagrams, we shall say that $T \in L(\overline{X}, \overline{Y})$ is of finite rank if T factors through a diagram

$\overline{F}\in\mathcal{F}$.

We note that since for $\overline{F}\in\mathcal{F}$, $L(\overline{X},\overline{F}) = L(\overline{F}',\overline{X}')$, it is sufficient for the study of operators of finite rank to analyze the operators having as domain finite dimensional doolittle diagrams. Thus, combining Corollary 2.4 with the description of operators of rank 1 above, we have proved the following results.

3.3. Proposition. Let $\overline{F}\in\mathcal{F}$, $\overline{Y}\in\mathcal{B}$ and let $T\in L(\overline{F},\overline{Y})$. In terms of the dual basis for $\overline{F}'$ described in Section 2, we have

$$T_i = \sum_{j=1}^{k_i} k'_{ij}\otimes x_{ij} + \sum_{m=1}^{d} \sigma'_i(d'_m)\otimes y_{im} + \sum_{n=1}^{p_i} \sigma'_i(p'_{in})\otimes z_{in},$$

where $x_{ij}\in K_i\overline{Y}$, $y_{im}\in \mathrm{im}(\delta_i)$, and $z_{in}\in Y_i$. Furthermore, there exists $y_m\in \Delta\overline{Y}$ such that $y_{im} = \delta_i(y_m)$.

3.4. Proposition. Let $T\in L(\overline{X},\overline{Y})$ be of finite rank. Then there exist operators T_I, T_{K_0}, T_{K_1}, T_{P_0}, and T_{P_1} such that

$$T = T_I + T_{K_0} + T_{K_1} + T_{P_0} + T_{P_1}$$

and such that

$$T_I \in \Delta\overline{X}'\otimes\Delta\overline{Y},$$
$$T_{K_i} \in X'_i\otimes K_i\overline{Y}, \text{ and}$$
$$T_{P_i} \in K_i\overline{X}'\otimes Y_i.$$

3.5. Remark. We point out that although the representation in Proposition 3.3 is unique (given the basis), the decomposition in 3.4 is not.

4. Applications.

There are several reasons for the importance of the study of finite dimensional doolittle diagrams in our theory of interpolation. One of these is the fact that we have an analogue to the property in $\mathfrak{B}$ that each Banach space is the inductive limit of its finite dimensional subspaces. Another is that we can often determine the behaviour of a functor on $\overline{\mathfrak{B}}$ by merely looking at its behaviour on elements of $\overline{\mathcal{F}}$, a topic which is discussed in the next chapter under the notion of "computability".

First we state our analogue of the result on inductive limits of finite dimensional spaces.

4.1. Proposition. Let $\overline{X}\in\overline{\mathfrak{B}}$. Then there exists $\{\overline{F}_\alpha\in\overline{\mathcal{F}}\mid\alpha\in A\}$, where A is some directed set, such that

(i) $\overline{F}_\alpha$ is a subdiagram of $\overline{X}$ for each $\alpha\in A$, and

(ii) $\overline{X} = \varinjlim \overline{F}_\alpha$ in the sense that $\Delta\overline{X} = \varinjlim \Delta\overline{F}_\alpha$, $X_0 = \varinjlim F_{\alpha,0}$, $X_1 = \varinjlim F_{\alpha,1}$, and $\Sigma\overline{X} = \varinjlim \Sigma\overline{F}_\alpha$.

4.2. Lemma. Let $\overline{E}$ and $\overline{F}$ be subdiagrams of $\overline{X}$. Let $\overline{G} = \overline{E}+\overline{F}$ be defined by taking G_i to be the sum space E_i+F_i of the subspaces E_i and F_i of X_i, $\Delta\overline{G}$ to be the pullback of

$$\begin{array}{ccc} & & G_0 \\ & & \downarrow \\ G_1 & \longrightarrow & \Sigma\overline{X}, \end{array}$$

and $\Sigma\overline{G}$ to be the pushout of

$$\begin{array}{ccc} \Delta\overline{G} & \longrightarrow & G_0 \\ \downarrow & & \\ G_1 & & . \end{array}$$

Then $\overline{G}$ is a subdiagram of $\overline{X}$ and there are $\mathfrak{B}$-morphisms

$$\begin{array}{c} \overline{E} \searrow \\ \overline{F} \nearrow \end{array} \overline{G} \longrightarrow \overline{X} \, .$$

<u>Proof</u>: The result follows directly from the definition of a subdiagram since clearly $G_i \to X_i$ are isometries, i=0,1, and $\Sigma\overline{G} \to \Sigma\overline{X}$ is a monomorphism. □

<u>Proof of Proposition 4.1</u>: Let $F_0 \subset X_0$ and $F_1 \subset X_1$ be finite dimensional subspaces (i.e. isometrically embedded). For $\alpha = (F_0, F_1)$, we define $\Delta\overline{F}_\alpha$ to be the pullback of

$$\begin{array}{ccc} & & F_0 \\ & & \downarrow \\ F_1 & \longrightarrow & \Sigma\overline{X} \end{array}$$

and $\Sigma\overline{F}_\alpha$ to be the pushout of

$$\begin{array}{ccc} \Delta\overline{F}_\alpha & \longrightarrow & F_0 \\ \downarrow & & \\ F_1 & & . \end{array}$$

The set A of all $\alpha = (F_0, F_1)$ is a set directed by inclusion. $\{\overline{F}_\alpha\}$

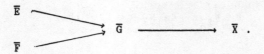

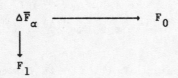

is a directed system of subdiagrams of $\overline{X}$ by Lemma 4.2, and for $\alpha<\beta$, we have a $\mathfrak{B}$-monomorphism $\overline{F}_\alpha \to \overline{F}_\beta$. Clearly, $\underrightarrow{\lim}\, F_{\alpha,i} = X_i$, $i=0,1$. If $x\in\Delta\overline{X}$, then letting F_i be the subspace of X_i spanned by $\delta_i(x)$, we have $x\in\Delta\overline{F}_\alpha$, where $\alpha = (F_0,F_1)$, so $\underrightarrow{\lim}\, \Delta\overline{F}_\alpha = \Delta\overline{X}$. Thus, it remains to prove that $\underrightarrow{\lim}\, \Sigma\overline{F}_\alpha = \Sigma\overline{X}$. By definition there exists $u:\underrightarrow{\lim}\, \Sigma\overline{F}_\alpha \to \Sigma\overline{X}$. First, we can see that u is a quotient map since if $\varepsilon>0$ and $x\in\Sigma\overline{X}$, we may choose $x_0\in X_0$ and $x_1\in X_1$ such that $\sigma_0 x_0 + \sigma_1 x_1 = x$ and

$$(*) \qquad \|x_0\|_{X_0} + \|x_1\|_{X_1} < (1+\varepsilon)\|x\|,$$

and define F_i to be the subspace of X_i spanned by x_i, $i=0,1$, respectively. Finally, to show that u is injective, we shall use duality and prove instead that

$$u':(\Sigma\overline{X})' = \Delta\overline{X}' \to (\underrightarrow{\lim}\, \Sigma\overline{F}_\alpha)' = \underleftarrow{\lim}\, \Delta\overline{F}'_\alpha$$

is surjective. We note that if $\alpha<\beta$, then $i_{\alpha,\beta}:\Sigma\overline{F}_\alpha \to \Sigma\overline{F}_\beta$ and $p_{\alpha,\beta} = (i_{\alpha,\beta})':\Delta\overline{F}'_\beta \to \Delta\overline{F}'_\alpha$. Thus, an element of $\underleftarrow{\lim}\, \Delta\overline{F}'_\alpha$ is a net $\{f_\alpha\}\in\Delta\overline{F}'_\alpha$ such that $f_\beta(i_{\alpha,\beta}(x)) = f_\alpha(x)$ for $x\in\Sigma\overline{F}_\alpha$. If $\{f_\alpha\}\in\underleftarrow{\lim}\, \Delta\overline{F}'_\alpha$, then for every $x\in\Sigma\overline{F}_\alpha$, $|f_\alpha(x)| \le C\|u(x)\|$ by (*). Hence, $\{f_\alpha\}$ defines an element $f\in(\Sigma\overline{X})' = \Delta\overline{X}'$, which completes our proof that u' is surjective. □

4.3. <u>Corollary</u>. If $\overline{X}\varepsilon\mathfrak{BC}$, then

$$\overline{X} = \underrightarrow{\lim}\{\overline{F}_\alpha \in \overline{\mathcal{E}} \,|\, \alpha \in A\},$$

where $\overline{\mathcal{E}}$ denotes the category of all finite dimensional Banach couples.

<u>Proof</u>: We recall that a subdiagram of a Banach couple is again a

Banach couple. □

4.4. Remark. If $\overline{X}$ is a regular Banach couple, then we can express $\overline{X}$ as an inductive limit of regular subcouples by verifying that

$$\overline{X} = \underrightarrow{\lim}\{\overline{F}_\alpha \in \overline{\mathcal{E}} \mid \alpha \in A\} = \underrightarrow{\lim}\{\overline{F}^o_\alpha \in \overline{\mathcal{E}} \mid \alpha \in \overline{A}\},$$

where $\overline{F}^o$ is defined as in I.2 to be the diagram

$$\begin{array}{ccc} \Delta\overline{F} & \longrightarrow & F_0^o \\ \downarrow & & \downarrow \\ F_1^o & \longrightarrow & \Sigma(\overline{F}^o). \end{array}$$

(F_i^o is the closure of the image of δ_i in F_i.)

In Banach spaces one has an isomorphism of $L(E,X)'$ and $E\otimes X'$ when E is finite dimensional. In $\overline{\mathfrak{B}}$ this isomorphism does not hold since if we take, for example, the diagram $\overline{E}=\overline{K}_0$, we obtain by 1.3,

$$L(\overline{K}_0,\overline{X})' = (K_0\overline{X})' = X_0'/(\ker\sigma_0)^{\perp}$$

and

$$\overline{K}_0\otimes\overline{X}' = X_0'/\Delta,$$

which need not be the same. However, it is important for our work on duality to know when $L(\overline{E},\overline{X})'$ and $\overline{E}\otimes\overline{X}'$ agree.

We start by describing the space $\overline{E}\otimes\overline{X}$, $\overline{E}\in\overline{\mathcal{F}}$, using the basis for $\overline{E}$ as described in Proposition 2.3. Thus, $\{k_{ij}\}_{j=1}^{k_i}$, $\{d_j\}_{j=1}^{d}$, are vectors in $\Delta\overline{E}$ and $\{p_{ij}\}_{j=1}^{p_i}$ are vectors in E_i such that

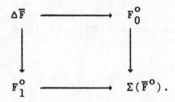

$$\{\delta_i(k_{ij})\}_{j=1}^{k_i} \cup \{\delta_i(d_j)\}_{j=1}^{d} \cup \{p_{ij}\}_{j=1}^{p_i}$$

is a basis for E_i, $i=0$ and 1.

Now $\overline{E}\otimes\overline{X}$ is the pushout of the diagram

$$\begin{array}{ccc} \Delta\overline{E}\otimes\Delta\overline{X} & \longrightarrow & E_0\otimes X_0 \\ \big\downarrow & & \\ E_1\otimes X_1 & & \end{array},$$

so it is a quotient space of $E_0\otimes X_0 \amalg E_1\otimes X_1$. If $t \in \overline{E}\otimes\overline{X}$, we start by choosing $t_0 \in E_0\otimes X_0$ and $t_1 \in E_1\otimes X_1$ so that $t_0 + t_1 = t$ in $\overline{E}\otimes\overline{X}$. Now we can write

$$t_i = \sum_{j=1}^{k_i} \delta_i(k_{ij})\otimes x_{ij} + \sum_{m=1}^{d} \delta_i(d_{im})\otimes y_{im} + \sum_{n=1}^{p_i} p_{in}\otimes z_{in}.$$

Hence,

$$t = \sum_{i=0}^{1} \sum_{j=1}^{k_0} k_{ij}\otimes\xi_{ij} + \sum_{m=1}^{d} d_m\otimes y_{\sigma m} + \sum_{i=0}^{1} \sum_{n=1}^{p_i} p_{in}\otimes z_{in}$$

where $\xi_{ij} \in X_i/\mathcal{C}\ell(\mathrm{im}(\delta_i))$, $y_{\sigma m} \in \Sigma\overline{X}$, and $z_{in} \in X_i$.

4.5. <u>Proposition</u>. Let $\overline{E}\in\mathscr{F}$ and let $\overline{X}\in\mathscr{B}$. Then the dual of $L(\overline{E},\overline{X})$ is $\overline{E}\otimes\overline{X}'$ if either

(i) $\overline{X}$ is reflexive or

(ii) $\overline{E}$ is a Banach couple, i.e. if $\overline{E}\in\overline{\mathscr{C}}$.

<u>Proof</u>: $L(\overline{E},\overline{X})$ is defined as the pullback of the diagram

$$\begin{array}{ccc} & & L(E_0,X_0) \\ & & \downarrow \\ L(E_1,X_1) & \longrightarrow & L(\Delta\overline{E},\Sigma\overline{X}), \end{array}$$

so in particular we have an isometric embedding

$$L(\overline{E},\overline{X}) \to L(E_0,X_0)\pi L(E_1,X_1).$$

Therefore, $L(\overline{E},\overline{X})'$ is a quotient of

$$(L(E_0,X_0)\pi L(E_1,X_1))' = L(E_0,X_0)'\amalg L(E_1,X_1)'.$$

Using the fact that $L(E,X)' = E\otimes X'$ when E is a finite dimensional Banach space, we may express $L(\overline{E},\overline{X})'$ as

$$(E_0\otimes X_0'\amalg E_1\otimes X_1')/L(\overline{E},\overline{X})^{\perp}.$$

Also, since $(\overline{E}\otimes\overline{X}')' = L(\overline{E},\overline{X}'')$ by adjointness, we may write $\overline{E}\otimes\overline{X}'$ as

$$(E_0\otimes X_0'\amalg E_1\otimes X_1')/L(\overline{E},\overline{X}'')^{\perp}.$$

Since we obviously have $L(\overline{E},\overline{X}) \subset L(\overline{E},\overline{X}'')$, it follows that $L(\overline{E},\overline{X}'')^{\perp} \subset L(\overline{E},\overline{X})^{\perp}$. Thus, there is a quotient map

$$q:\overline{E}\otimes\overline{X}' \to L(\overline{E},\overline{X})'.$$

This map will be an isomorphism if for every non-zero $t \in \overline{E}\otimes\overline{X}'$ there exists $T \in L(\overline{E},\overline{X})$ such that $\langle qt, T\rangle \neq 0$. Using the description of $\overline{E}\otimes\overline{X}'$ given above, such a t has the form

$$t = \sum_{i=0}^{1} \sum_{i=1}^{k_i} k_{ij} \otimes \xi'_{ij} + \sum_{m=1}^{d} d_m \otimes y'_{\sigma m} + \sum_{i=0}^{1} \sum_{i=1}^{p_i} p_{in} \otimes z'_{in}$$

for $\xi'_{ij} \in X'_i/\mathcal{C}\ell(\text{im } \sigma'_i)$, $y'_{\sigma m} \in \Sigma\overline{X}'$, $z'_{in} \in X'_i$, and $t = (t_0,t_1) + N$, where $N = L(\overline{E},\overline{X}'')^{\perp}$ and

$$t_i = \sum_{j=1}^{k_i} \delta_i(k_{ij}) \otimes x'_{ij} + \sum_{m=1}^{d} \delta_i(d_m) \otimes y'_{im} + \sum_{n=1}^{p_i} p_{in} \otimes z'_{in},$$

where x'_{ij}, y'_{im}, $z'_{in} \in X'_i$. First assume that for some m, $\delta'_0 y'_{0m} + \delta'_1 y'_{1m} = y'_{\sigma m} \neq 0$. Then we choose $x \in \Delta\overline{X}$ such that $\langle y'_{\sigma m}, x\rangle = 1$ and define $T = (T_0,T_1)$ to be $(\sigma'_0(d'_m) \otimes \delta_0 x, \sigma'_1(d'_m) \otimes d_1 x)$. Then $\langle qt, T\rangle = 1$. If $y'_{\sigma m} = 0$ for $m=1,\ldots,d$, then we assume next that $z'_{in} \neq 0$ for some n and some i, say $z'_{0n} \neq 0$. Then we choose $x_{0n} \in X_0$ such that $\langle z'_{0n}, x_{0n}\rangle = 1$ and let $T = (T_0,T_1) = (p'_{0n} \otimes x_{0n}, 0)$ and verify that $\langle qt, T\rangle = 1$. Hence, the only case remaining to consider is when all $y'_{\sigma m} = 0$ and all $z'_{in} = 0$ but some $\xi'_{ij} \neq 0$, say ξ'_{0j}. Then we may choose any $x_0 \in K_0\overline{X}$ and define $T = (T_0,T_1)$ to be $(k'_{ij} \otimes x_0, 0)$. It follows that $\langle qt, T\rangle = \langle \xi'_{0j}, x_0\rangle$. Now if $\overline{X}$ is reflexive, then $\mathcal{C}\ell(\text{im } \sigma'_0) = (K_0\overline{X})^{\perp}$, so since $\xi'_{0j} \notin \mathcal{C}\ell(\text{im } \sigma'_0)$, there exists $x_0 \in K_0\overline{X}$ such that $\langle \xi'_{0j}, x_0\rangle \neq 0$. If $\overline{X}$ is non-reflexive, then there can be some $\xi' \in (K_0\overline{X})^{\perp}$ with $\xi' \notin \mathcal{C}\ell(\text{im } \sigma'_0)$ and then the element $t = (t_0,t_1) = (\delta_0(k) \otimes \xi', 0)$, where k is any element in the subspace of $\Delta\overline{E}$ determined by $\{k_{0j}\}$, has the property that $\langle qt, T\rangle = 0$ for all $T \in L(\overline{E},\overline{X})$, so q is non-injective. Hence, we must conclude that if $\overline{X}$ is non-reflexive, the theorem is true only when the first two cases occur. The condition that $\overline{E}$ is a couple is sufficient to insure this since the basis for $\overline{E}$ would then contain no elements of the form k_{ij}. □

CHAPTER VI

KAN EXTENSIONS

1. Definition.

It is a common experience that a functor has a natural construction only for a subcategory of the one under study and that it must then be extended to the whole category. This usually gives rise to what in category theory is called a left or right Kan extension. The left Kan extension is a minimal extension, while the right Kan extension is maximal.

In general the situation we are faced with is obtaining a left or right $\mathcal{C}$-adjoint to the restriction functor

$$U:\mathcal{C}^{\mathcal{D}} \to \mathcal{C}^{\mathcal{A}},$$

where $\mathcal{A}$ is a full subcategory of the $\mathcal{C}$-category $\mathcal{D}$ ($\mathcal{C}$ a closed category as in Chapter IV), and where $\mathcal{C}^{(-)}$ denotes the category of $\mathcal{C}$-functors from $(-)$ to $\mathcal{C}$. If K denotes the inclusion of $\mathcal{A}$ in $\mathcal{D}$, then the left adjoint of U, Lan_K, is called the <u>left Kan extension along K</u>, and the right adjoint, Ran_K, is called the <u>right Kan extension along K</u>. Explicitly, given $F:\mathcal{A} \to \mathcal{C}$, then the following diagram is commutative

$$\begin{array}{ccc} \mathcal{C} & & \\ F\big\downarrow & \quad Lan_K(F) & \\ \mathcal{A} & \longrightarrow & \mathcal{D}\,. \end{array}$$

Furthermore, for every $G:\mathcal{D} \to \mathcal{C}$ there is an isomorphism (natural in F and G)

$$NAT(Lan_K F,\ G) \simeq NAT(F,\ G \circ K),$$

which, of course, says that $Lan_K(F)$ is the minimal extension of F from $\mathcal{A}$ to $\mathcal{D}$ in the sense that if $G:\mathcal{D} \to \mathcal{C}$ is any functor agreeing with F on $\mathcal{A}$, i.e. such that $G \circ K = F$, then there is a natural transformation from $Lan_K F$ to G extending the identity on $\mathcal{A}$. The analogous description can be given for $Ran_K F$, which is a maximal extension in the above sense.

In the remainder of this chapter we shall study Kan extensions that arise naturally in the context of Banach spaces and interpolation theory. In each case we shall give a concrete description of the functor $Lan_K F$ so tersely described here. For additional information concerning Kan extensions we recommend [18] for a general discussion and [5] for a discussion oriented to the category of Banach spaces.

2. Examples.

Our concern in this work lies with $\mathcal{C}=\mathcal{B}$ and $\mathcal{D}=\mathcal{B}$, $\overline{\mathcal{B}\mathcal{C}}$, or $\overline{\mathcal{B}}$. Several examples of Kan extensions in this context are of interest and some will be of importance in the theory of dual functors and interpolation functors.

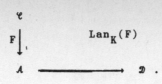

2.1. Example. Let $\mathcal{A}$ be the subcategory of $\mathcal{B}$ consisting of only I. Then a functor from $\mathcal{A}$ to $\mathcal{B}$ may be identified with a Banach space A. The left Kan extension of this functor, call it Lan_A, is characterized by the property that it is minimal with respect to functors F such that FI = A. Now bearing in mind that $A\otimes-$ is a functor taking I to A and recalling the counit map

$$\varepsilon: L(I,X)\otimes FI = X\otimes FI \to FX$$

described in IV.4.1 in the $\overline{\mathcal{B}}$-setting, we conclude immediately that $A\otimes-$ is the minimal functor we are seeking, i.e. that $\mathrm{Lan}_A = A\otimes$.

Similarly, Ran_A is characterized by being maximal among functors F such that FI = A. Suppose that F is such a functor. If $f \in FX$ and $x' \in L(X,I) = X'$, then $F(x')(f) \in A$, so f may be thought of as a map from X' to A, which it may be checked is actually a $\mathcal{B}$-morphism. Since $L(I',A) = A$, we may conclude that $\mathrm{Ran}_A X = L(X',A)$.

It is worth pointing out that when A=I, Lan_A is the identity functor but that $\mathrm{Ran}_A X = X''$. This illustrates a phenomenon that occurs frequently in practice and may be summarized by saying that the right Kan extension is "too big". We will encounter similar instances of this below.

2.2. Example. If $\overline{\mathcal{A}}$ is the subcategory of $\overline{\mathcal{B}}$ consisting of a single element $\overline{A}$, then a functor F from $\overline{\mathcal{A}}$ to $\mathcal{B}$ may be identified with an $L\overline{A}$-module since if $T:\overline{A} \to \overline{A}$, then $FT:F\overline{A} \to F\overline{A}$ gives us a module action. Letting $F\overline{A} = A$, we denote the left and right Kan extensions of F along the inclusion of $\overline{\mathcal{A}}$ to $\overline{\mathcal{B}}$ simply by Lan_A and Ran_A, respectively. Then we may observe that

$$\mathrm{Lan}_A\overline{X} = L(\overline{A},\overline{X})\otimes_{L\overline{A}}A$$

and

$$\mathrm{Ran}_A\overline{X} = L_{L\overline{A}}(L(\overline{X},\overline{A}),A),$$

where $\otimes_{L\overline{A}}$ and $L_{L\overline{A}}$ denote, respectively, the Banach module tensor product over $L\overline{A}$ and the $L\overline{A}$-module linear maps (see [12]). That $L\overline{A}\otimes_{L\overline{A}}A = A = L_{L\overline{A}}(L\overline{A},A)$ follows from general principles since $L\overline{A}$ is a unital Banach algebra. Moreover, if $G:\overline{\mathfrak{B}} \to \mathfrak{B}$ is any functor such that $G\overline{A} = A$, then natural maps ξ and η,

$$L(\overline{A},\overline{X})\otimes_{L\overline{A}}A \xrightarrow{\ \xi\ } G\overline{X} \xrightarrow{\ \eta\ } L_{L\overline{A}}(L(\overline{X},\overline{A}),A),$$

can be defined by $\xi(T\otimes a) = GT(a)$ and $\eta(x)(S) = GS(x)$, showing that $L(\overline{A},-)\otimes_{L\overline{A}}A$ and $L_{L\overline{A}}(L(-,\overline{A}),A)$ are indeed the minimal and maximal such functors.

We remark that it follows from the definition of the left Kan extension that for any $G:\overline{\mathfrak{B}} \to \mathfrak{B}$, $\mathrm{NAT}(\mathrm{Lan}_A,G)$ is determined by the single diagram $\overline{A}$ since

$$\mathrm{NAT}(\mathrm{Lan}_A,G) = \mathrm{NAT}(F,G|_{\overline{\mathcal{A}}}) = L_{L\overline{A}}(A,G\overline{A}).$$

Hence, the above natural transformation $\xi:\mathrm{Lan}_A \to G$ is the map corresponding to the identity map on A. A similar observation holds for $\mathrm{NAT}(G,\mathrm{Ran}_A)$.

The following proposition, which will be useful later on, is a simple consequence of the definition of Lan_A.

2.3. Proposition. If A and B are $L\overline{A}$-modules with A dense in B, then $\mathrm{Lan}_A\overline{X}$ is dense in $\mathrm{Lan}_B\overline{X}$ for every $\overline{X}\in\overline{\mathfrak{B}}$.

As a simple special case of the preceding situation, let $\overline{A}=\overline{I}$

and A=I. Then

$$\mathrm{Lan}_A\overline{X} = L(\overline{I},\overline{X})\otimes_{L\overline{I}} I = L(\overline{I},\overline{X}) = \Delta\overline{X}$$

and

$$\mathrm{Ran}_A\overline{X} = L_{L\overline{I}}(L(\overline{X},\overline{I}),I) = L(\overline{X},\overline{I})' = L(\overline{I},\overline{X}')' = (\Delta\overline{X}')' = \Sigma\overline{X}'' = (\Sigma\overline{X})''.$$

This example also shows, as did Example 2.1, that the left Kan extension is a useful functor while the right Kan extension is too big.

If $\overline{A} = \overline{I}(s,t)$, as defined by V.1.1, and if A=I, then $\mathrm{Lan}_A\overline{X} = \Delta(s,t,\overline{X})$ while $\mathrm{Ran}_A\overline{X} = \Sigma(s,t,\overline{X}'') = \Sigma(s,t,\overline{X})''$.

Two other special cases of interest are $\overline{A} = \overline{P}_i$ and $\overline{A} = \overline{K}_i$, i=0,1. Recalling Proposition V.1.4, we see that if $\overline{A} = \overline{P}_i$ and A=I, then

$$\mathrm{Lan}_A\overline{X} = L(\overline{P}_i,\overline{X}) = X_i,$$

while

$$\mathrm{Ran}_A\overline{X} = L(\overline{X},\overline{P}_i)' = L(\overline{K}_i,\overline{X}')' = X_i''/\mathcal{C}\ell(\mathrm{im}\delta_i).$$

If $\overline{A} = \overline{K}_i$ and A=I, then we have

$$\mathrm{Lan}_A\overline{X} = L(\overline{K}_i,\overline{X}) = K_i\overline{X}$$

and

$$\mathrm{Ran}_A\overline{X} = L(\overline{X},\overline{K}_i)' = L(\overline{P}_i,\overline{X}')' = X_i''.$$

2.4. Example. If $\bar{A}$ is non-trivial, then we shall see that $\mathrm{Lan}_{\Delta\bar{A}}\bar{X}$ is topologically isomorphic to $\Delta\bar{X}$, actually $\mathcal{B}$-isomorphic if $\bar{A}$ is unital (as defined in IV.2). We shall prove the latter result, since the proof of the former follows a similar pattern. If $\bar{A}$ is unital with unit u and dual unit u', then we shall follow the notation of IV.5.2 in writing $\hat{x}$ for the operator $u'\otimes x:\bar{A} \to \bar{X}$ defined for $x\in\Delta\bar{X}$ by $(u'\otimes x)(a) = \langle u',ja\rangle x$, for $a\in\Delta\bar{A}$.

2.5. Proposition. Let $\bar{A}$ be unital. Then $L(\bar{A},\bar{X})\otimes_{L\bar{A}}\Delta\bar{A} = \Delta\bar{X}$ for all $\bar{X}\in\bar{\mathcal{B}}$.

Proof: Let $\Sigma T_i\otimes a_i \in L(\bar{A},\bar{X})\otimes_{L\bar{A}}\Delta\bar{A}$. Using the property of the Banach module tensor product, we have

$$T_i\otimes a_i = T_i\otimes\hat{a}_i(u) = T_i\circ\hat{a}_i\otimes u = T_i(a_i)\otimes u.$$

Hence,

$$\Sigma T_i\otimes a_i = \Sigma T_i(a_i)\otimes u,$$

and clearly $\Sigma T_i a_i \in \Delta\bar{X}$. Moreover, it is easy to check that the norms coincide. □

2.6. Example. If we are given a classical interpolation functor F from $\bar{\mathcal{B}}\mathcal{C}$ to $\mathcal{B}$, the Kan extension construction would be one way of extending F to the $\bar{\mathcal{B}}$-setting. In fact, both the left and right Kan extensions are automatically quasi-interpolation functors since the constant diagrams belong to $\bar{\mathcal{B}}\mathcal{C}$. In Chapter IX we see by means of another Kan extension construction that our extensions of the classical J-method and the complex C_θ-method are actually obtained by left Kan extensions.

3. Computable functors.

A simple but important situation involving Kan extensions arises when F is a functor defined only on the subcategory of finite dimensional Banach spaces $\mathcal{F}$. In this case the left Kan extension is given by

$$\mathrm{Lan}_K FX = \varinjlim\{FY \mid Y \subseteq X,\ Y \in \mathcal{F}\}.$$

Clearly, the restriction $U\mathrm{Lan}_K F$ is still F, but $\mathrm{Lan}_K UF$, for $F:\mathcal{B} \to \mathcal{B}$, is not necessarily equal to F. We call F computable, adopting the now standard terminology of [10], when $\mathrm{Lan}_K UF = F$. We remark that the class of computable functors on $\mathcal{B}$ includes the functors $L_p(\mu;-)$, $1 \le p < \infty$, where μ denotes a fixed measure space, $C(\Omega,-)$, where Ω denotes a compact Hausdorff space, and $X\otimes_\beta -$, where $\otimes_\beta$ denotes a tensor norm in the sense of Grothendieck (see [22]).

It will be important to our work on dual functors and interpolation functors to have a notion of computability for functors from $\overline{\mathcal{B}}$ to $\mathcal{B}$. To this end we shall denote by $\overline{\mathcal{F}}$ the full subcategory of $\overline{\mathcal{B}}$ consisting of finite dimensional doolittle diagrams as defined in Chapter V and by $\overline{\mathcal{E}}$ the subcategory of finite dimensional couples. We recall that $\overline{E}\in\overline{\mathcal{F}}$ or $\overline{\mathcal{E}}$ is called a finite dimensional subdiagram of $\overline{X}$ (we write $\overline{E} \subseteq \overline{X}$) if there is an extremal monomorphism in $\overline{\mathcal{B}}$ from $\overline{E}$ to $\overline{X}$, which means that $\Delta\overline{E}$, E_0, and E_1 are subspaces of $\Delta\overline{X}$, X_0, and X_1, respectively. It was noted in IV.3.8 that the subdiagram of a Banach couple is again a Banach couple. Moreover, we recall from V.4.1 that every $\overline{X}\in\overline{\mathcal{B}}$ ($\overline{X}\in\overline{\mathcal{B}\mathcal{C}}$) is in a strong sense the direct limit of its finite dimensional subdiagrams (subcouples). As in the Banach space case we can form the left Kan extensions along the inclusion K of $\overline{\mathcal{E}}$ or $\overline{\mathcal{F}}$ in $\overline{\mathcal{B}}$. However, it turns out in view of the duality theorem we shall prove in the next chapter and the discussion there that the left Kan extension along $K:\overline{\mathcal{E}} \to \overline{\mathcal{B}}$ is the more useful functor. Thus, we shall speak of both $\mathcal{F}$- and $\mathcal{E}$-computability although the latter notion will be more dominant.

In the case that UF is a Δ-functor, we have for any $\overline{E}\in\overline{\mathcal{E}}$, $F\overline{E} \approx \Delta\overline{E}$ and $L(\overline{E},\overline{X})\otimes_{L\overline{E}}F\overline{E} \approx L(\overline{E},\overline{X})\otimes_{L\overline{E}}\overline{E} \approx \Delta\overline{X}$, so $\mathrm{Lan}_K UF\overline{X} = \underrightarrow{\lim}\ L(\overline{E},\overline{X})\otimes_{L\overline{E}}F\overline{E}$ is the completion of $\Delta\overline{X}$ with respect to the smaller norm $\|\ \|_L$, where

$$\|x\|_L = \inf\{\Sigma\|T_i\|\|e_i\| \mid x = \Sigma T_i e_i,\ T_i\in L(\overline{E}_i,\overline{X}),\ e_i\in F\overline{E}_i,\ \overline{E}_i\in \overline{\mathcal{E}}\}.$$

4. Aronszajn-Gagliardo functors.

An important theorem of interpolation theory due to Aronszajn and Gagliardo [1] states that any (classical) interpolation space A for the Banach couple (A_0,A_1) may be obtained by an interpolation method - or functor - applied to (A_0,A_1). This categorical-sounding result really says that interpolation spaces do not appear by chance but rather by means of a process. The fact that Aronszajn and Gagliardo actually construct interpolation functors F and H on the category of Banach couples which are minimal and maximal with respect to the property $F(A_0,A_1) = A = H(A_0,A_1)$ leads us to suspect the presence of a Kan extension construction. Thus, we follow the model of Example 2.2 to obtain a $\overline{\mathcal{B}}$-setting analogue of the Aronszajn-Gagliardo functor.

We begin quite generally by choosing $\overline{A}\in\overline{\mathcal{B}}$ and A any $L\overline{A}$-module. Thus, as was previously remarked, A may be identified with a functor $\{\overline{A}\} \to \mathcal{B}$. Then $\mathrm{Lan}_A\overline{X} = L(\overline{A},\overline{X})\otimes_{L\overline{A}}A$ and $\mathrm{Ran}_A\overline{X} = L_{L\overline{A}}(L(\overline{X},\overline{A}),A)$ are the minimal and maximal functors, respectively, mapping $\overline{A}$ to A. These functors closely resemble the classical Aronszajn-Gagliardo functors F and H, the definitions of which are given as follows:

$$F\overline{X} = \{x \in \Sigma\overline{X} \mid x = \Sigma T_j a_j,\ T_j \in L(\overline{A},\overline{X}),\ a_j \in A\}$$
$$H\overline{X} = \{x \in \Sigma\overline{X} \mid Tx \in A \text{ for all } T \in L(\overline{X},\overline{A})\}.$$

In fact, in all cases known to us F is just Lan_A when A is an interpolation space in the traditional setting. On the other hand, H is the necessary improvement on the right Kan extension, which we have seen is usually too big. We have observed that the left Kan extension behaves especially well for interpolation in $\overline{\mathcal{B}}$ when A is taken to be a Δ-interpolation space and the right Kan extension (suitably modified) when A is a Σ-interpolation space. The following results will confirm this observation and justify our future restriction of these constructions to these cases.

4.1. <u>Proposition</u>. Let $\overline{A}$ be non-trivial and let A be a Δ-interpolation space for $\overline{A}$. Then Lan_A is a Δ-interpolation functor and is minimal among Δ-interpolation functors mapping $\overline{A}$ to A.

<u>Proof</u>: Assume $\overline{A}$ is unital. Then by Proposition 2.5, $\mathrm{Lan}_{\Delta\overline{A}}\overline{X} = \Delta\overline{X}$. Hence, by Proposition 2.3, $\Delta\overline{X}$ is dense in $\mathrm{Lan}_A\overline{X}$ for each $\overline{X}$. From this fact it also follows that $\Delta JX = X$ is dense in $\mathrm{Lan}_A JX$ for each $X\in\mathcal{B}$. Since $\mathrm{Lan}_A JX$ is an $L(JX)$-module, we must have $\mathrm{Lan}_A JX = X$. This shows that Lan_A is a Δ-interpolation functor. Its minimality follows from the discussion in Example 2.2.

If $\overline{A}$ is not unital, the result follows by the same reasoning since $Lan_{\Delta\overline{A}}\overline{X}$ is $\overline{\mathcal{B}}_\infty$-isomorphic to $\Delta\overline{X}$. □

We point out that the condition of non-triviality on $\overline{A}$ is essential since if $\overline{A}=\overline{K}_i$ and $A=I$, $Lan_A\overline{X} = K_i(\overline{X})$ by 2.2, and, hence, $Lan_A JX = 0$ for all $X\in\mathcal{B}$.

4.2. Remark. We may apply the process $(\ )^s$ introduced in Chapter IV to Lan_A in order to obtain a Σ-interpolation functor. This involves, as we recall, taking the coimage of the map $\sigma: Lan_A\overline{X} \to \Sigma\overline{X}$. However, it is our point of view that it is undesirable to do this. In many cases, Lan_A is automatically a Σ-interpolation functor when A is a Σ-interpolation space, so everyone is happy. However, even when it is not, leaving the functor alone insures that its dual will have a good formal description, namely,

$$(Lan_A\overline{X})' = L_{L\overline{A}}(L(\overline{A},\overline{X}),A') = L_{L\overline{A}}(A,L(\overline{A},\overline{X})').$$

Pursuing this point further, if X is any Δ-interpolation space for $\overline{X}$, X' will be the subspace of $\Sigma\overline{X}'$ consisting of all x' such that $|\langle x',x\rangle| \le C\|x\|_X$, $x\in\Delta\overline{X}$, while $(X^s)'$ will be the weak* closure of $\Delta\overline{X}'$ in X', which is usually very difficult to work with. Thus, if we do not accept that X may be the norm completion of a smaller norm on $\Delta\overline{X}$ which is not necessarily a subspace of $\Sigma\overline{X}$, then we will have to accept weak*-completions of $\Delta\overline{X}'$. We consider the norm completions to be the better option.

Starting with the right Kan extension we can capture the idea of the maximal Aronszajn-Gagliardo functor by a suitable categorical construction. We define the functor $H_A\overline{X}$ to be the pullback of the diagram

$$\begin{array}{ccc} & & L_{L\overline{A}}(L(\overline{X},\overline{A}),A) \\ & & \Big\downarrow {\scriptstyle L_{L\overline{A}}(L(\overline{X},\overline{A}),\sigma)} \\ \Sigma\overline{X} & \xrightarrow[e]{} & L_{L\overline{A}}(L(\overline{X},\overline{A}),\Sigma\overline{A}), \end{array}$$

where e stands for the evaluation map $e(x)(T) = Tx$. If A is a Σ-interpolation space, which is the only case for which we will consider H_A, then $L_{L\overline{A}}(L(\overline{X},\overline{A}),\sigma)$ is mono, so that

$$H_A\overline{X} = \{(T,x) \in L_{L\overline{A}}(L(\overline{X},\overline{A}),A) \times \Sigma\overline{X} \mid L_{L\overline{A}}(L(\overline{X},\overline{A}),\sigma)(T) = e(x)\}$$

can be thought of as a subspace of $\Sigma\overline{X}$; alternatively, one can view $H_A\overline{X}$ as $\Sigma\overline{X} \cap L_{L\overline{A}}(L(\overline{X},\overline{A}),A)$.

4.3. <u>Proposition</u>. If A is a Σ-interpolation space for $\overline{A}$, then H_A is a Σ-interpolation functor and is the maximal such functor mapping $\overline{A}$ to A.

<u>Proof</u>: It is clear that $(H_A \circ J)X = X$, since in this case e can be viewed as a map

$$X \to L_{L\overline{A}}(L(X,\Delta\overline{A}),A).$$

Thus, H_A is trivially a Σ-interpolation functor. Since $L_{L\overline{A}}(L(\overline{A},\overline{A}),A) = A$, it is also clear that $H_A\overline{A} = A$.

To see that $H_A\overline{X}$ is the maximal such functor, let G be any Σ-interpolation functor such that $G\overline{A} = A$. Then $G\overline{X}$ is a subspace of $\Sigma\overline{X}$ such that for every $T \in L(\overline{X},\overline{A})$ $Tx \in A$ for $x \in G\overline{X}$; hence $G\overline{X} \subset H_A\overline{X}$. □

4.4. <u>Convention</u>. In view of 4.1 and 4.3 and our desire for relevance to the theory of interpolation in $\overline{\mathfrak{B}}$, it will be understood from this point on that A is a Δ-interpolation or a Σ-interpolation space when Lan_A or H_A, respectively, is discussed. The one exception to this convention will occur in Chapter IX.3.

We may now obtain the counterpart of Proposition 2.5.

4.5. Proposition. If $\overline{A}$ is unital, then $H_{\Sigma\overline{A}}\overline{X} = \Sigma\overline{X}$ for every $\overline{X}\in\overline{\mathfrak{B}}$.

Proof: We need only show that the norm-decreasing map e is actually an isometry. Let $x\in\Sigma\overline{X}$ and choose $x'\in(\Sigma\overline{X})'$ such that $\|x'\| = 1$ and $\langle x',x\rangle = \|x\|$. Then consider $\hat{x}' = u\otimes x':\overline{X} \to \overline{A}$, where u is the unit of $\overline{A}$. We have $\|\hat{x}'\| \le 1$ and $\|e(x)(\hat{x}')\| = \|\hat{x}'(x)\| = \|x\|$. □

5. Computability of Lan_A.

We investigate here the conditions under which Lan_A is $\overline{\mathcal{E}}$-computable since computability is a key to obtaining a duality theorem for the minimal method of interpolation theory. It is no surprise that a generalization of the familiar approximation property for Banach spaces arises as an important condition.

5.1. Definition. We say that $\overline{X}\in\overline{\mathfrak{B}}$ satisfies the metric approximation property if for every pair (C_0,C_1) of compact sets, $C_i\subset X_i$, and every $\varepsilon > 0$, there exists $P \in L(\overline{X},\overline{X})$ of finite rank (i.e. P_0,P_1 have finite rank) and norm bounded by 1 such that

$$\|P_i x-x\|_{X_i} < \varepsilon$$

for all $x \in C_i$, $i=0,1$.

Our result, then, is the following:

5.2. Proposition. Let $\bar{A}$ be a regular couple satisfying the metric approximation property. Then Lan_A is $\bar{\mathcal{E}}$-computable.

Proof: To show that Lan_A is $\bar{\mathcal{E}}$-computable, we shall take $T\otimes a \in L(\bar{A},\bar{X})\otimes_{L\bar{A}}A$, $\varepsilon > 0$, and find an operator $K \in L(\bar{A},\bar{X})$ factoring through a finite dimensional Banach couple such that

$$\|T\otimes a - K\otimes b\| < \|T\|\|\varepsilon\|,$$

for some $b\in\Delta\bar{A}$. Since $\Delta\bar{A}$ is dense in A, we may choose $b\in\Delta\bar{A}$ such that $\|a-b\|_A < \varepsilon/2$. Using now the approximation property and the regularity of $\bar{A}$, we find an operator $P\in L\bar{A}$ whose range lies in an element of $\bar{\mathcal{E}}$ such that

$$\|b-Pb\|_{\Delta\bar{A}} < \varepsilon/2.$$

We define $K = T\circ P$. Then we have

$$\begin{aligned}\|T\otimes a - K\otimes b\| &= \|T\otimes a - T\otimes b + T\otimes b - T\circ P\otimes b\| \\ &\le \|T\|\|a-b\|_A + \|T\|\|b-Pb\|_A \\ &\le \|T\|(\|a-b\|_A + \|b-Pb\|_{\Delta\bar{A}}) < \|T\|\varepsilon. \qquad \square\end{aligned}$$

CHAPTER VII

DUALITY

1. Dual Functors.

The notion of a dual functor was introduced by Fuks [9], and applied to the category $\mathfrak{B}$ by Mityagin and Švarc [19]. Linton [16] then made a serious study of dual functors and showed that it was possible to define dual functors not only for closed categories $\mathfrak{C}$, as the previous authors had done, but also for certain $\mathfrak{C}$-based categories.

The dual functor as defined by Mityagin and Švarc is closely associated to the duality of Banach spaces and the adjointness of the tensor product and hom functors. In fact, letting DF denote the dual functor of $F:\mathfrak{B} \to \mathfrak{B}$, DF is determined uniquely on objects by the following two conditions:

(i) $DL(X,-) = X\otimes -$,

(ii) $NAT(F,DG) = NAT(G,DF)$.

We note that, categorically speaking, condition (ii) says that D as an operation on the functor category $\mathfrak{B}^{\mathfrak{B}}$ is adjoint to itself on the right, much as the dual space construction is in view of the isomorphism

$$L(X,Y') = L(Y,X').$$

For any F satisfying (i) and (ii), we must have by the Yoneda lemma (IV.4.4)

$$DFX = NAT(L(X,-),DF) = NAT(F,X\otimes -),$$

which gives us the definition of the dual functor. Clearly, when $f \in \mathfrak{B}(X,Y)$ and $t \in NAT(F,X\otimes -)$, the definition

$$DF(f)(t) = (f\otimes -)\circ t$$

makes DF functorial on maps.

We immediately obtain the following desirable property:

1.1. Proposition. $D(X\otimes -) = L(X,-)$.

Proof: Clearly, given a natural transformation $t:X\otimes - \to Y\otimes -$, $t_I \in L(X,Y)$. Moreover, a map $f \in L(X,Y)$ gives rise to $f\otimes - = s$, which is obviously natural. While the statement $f\otimes 1_I = f = s_I$ needs no explanation, the claim that $s_I \otimes 1_Z = s_Z$ for all $Z \in \mathfrak{B}$ is at the crux of the meaning of naturality. The relation is obtained as follows: let $z \in Z$ and think of z as an element of $L(I,Z)$. Then naturality of s requires that the following diagram be commutative, which shows us that $s_Z(x\otimes z) = s_I(x)\otimes z$ for $x \in X$:

$$\begin{array}{ccc} X\otimes I & \xrightarrow{s_I} & Y\otimes I \\ {\scriptstyle 1_X\otimes z}\downarrow & & \downarrow{\scriptstyle 1_Y\otimes z} \\ X\otimes Z & \xrightarrow[s_Z]{} & Y\otimes Z. \end{array}$$

Hence, s is determined by s_I, which concludes our proof. □

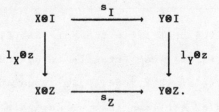

The general definition of the dual functor for functors $F:\mathcal{D} \to \mathcal{C}$, where $\mathcal{D}$ is a $\mathcal{C}$-based category, requires that there exist a symmetric bifunctor $\otimes:\mathcal{D} \times \mathcal{D} \to \mathcal{C}$ such that $-\otimes X$ has a (strong) right adjoint $\mathcal{C} \to \mathcal{D}$ for each $X \in \mathcal{D}$. When this is satisfied, the dual functor DF may be defined by

$$DFX = NAT(F, X\otimes -).$$

Other than the case $\mathcal{D} = \mathcal{C} = \mathcal{B}$ already discussed, we shall only be interested in $\mathcal{D} = \overline{\mathcal{B}}$. As we have seen by IV.2.2, $\overline{L}(\overline{X}, J(-)):\mathcal{B} \to \overline{\mathcal{B}}$ is the right adjoint of $\otimes\overline{X}:\overline{\mathcal{B}} \to \mathcal{B}$, so DF is well defined for $F:\overline{\mathcal{B}} \to \mathcal{B}$.

1.2. <u>Examples</u>. The theory developed above tells us in the context of $\overline{\mathcal{B}} = \mathcal{D}$ that $\overline{X}\otimes -$ and $L(\overline{X}, -)$ are dual to one another. There are several particular instances of this fact which are of interest. First, Σ and Δ are dual to one another since $\Sigma = \overline{I}\otimes -$ and $\Delta = L(\overline{I}, -)$. Moreover, $D\Delta(s,t,-) = \Sigma(s,t,-)$. On the other hand, the functor $(-)_i$ is self-dual since $X_i = \overline{P}_i\otimes\overline{X} = L(\overline{P}_i, \overline{X})$, $i=0,1$. Finally, $DK_i = DL(\overline{K}_i, -) = \overline{K}_i\otimes - = (-)_i/\Delta$.

2. <u>Descriptions of the Dual Functor</u>.

We have seen in IV.4 that often the set $NAT(F,G)$, as the projective limit of $L_{L\overline{X}}(F\overline{X}, G\overline{X})$, can actually be determined by a single diagram $\overline{X}$. In the case of $DF\overline{X} = NAT(F, \overline{X}\otimes -)$, we shall see in this section and the following sections that we may gain much information about $DF\overline{X}$ – and sometimes determine it – by specializing to a particular diagram $\overline{A}$ and considering the map

$$\gamma_{\overline{A}}:DF\overline{X} \to L_{L\overline{A}}(F\overline{A},\overline{X}\otimes\overline{A}),$$

where $\gamma_{\overline{A}}(t) = t_{\overline{A}}$. The first case in point is $\overline{A}=\overline{X}'$, where we see from the following result that we get some useful information about $DF\overline{X}$ with no restrictions on F.

2.1. Lemma. The map $\gamma_{\overline{X}'}:DF\overline{X} \to L_{L\overline{X}'}(F\overline{X}',\overline{X}\otimes\overline{X}')$ is an isometric inclusion for any $F:\overline{\mathfrak{B}} \to \mathfrak{B}$ and any $\overline{X}\in\mathfrak{B}$.

Proof: Let $t \in DF\overline{X}$. Then for every $\varepsilon>0$, there exists $\overline{Y}$ such that $\|t_{\overline{Y}}\| \geq (1-\varepsilon)\|t\|$. Now we can choose $y \in F\overline{Y}$ such that $\|y\| = 1$ and $\|t_{\overline{Y}}(y)\| \geq (1-\varepsilon)\|t_{\overline{Y}}\|$. Then since $t_{\overline{Y}}(y) \in \overline{X}\otimes\overline{Y}$, there exists $T \in (\overline{X}\otimes\overline{Y})' = L(\overline{Y},\overline{X}')$, $\|T\| = 1$, such that

$$\langle T,t_{\overline{Y}}(y)\rangle = Tr(t_{\overline{X}'}(FT(y)) = \|t_{\overline{Y}}(y)\|.$$

But then $x' = F(T)(y) \in F(\overline{X}')$ and $\|x'\| \leq 1$, so

$$\|t_{\overline{X}'}\| \geq \|t_{\overline{X}'}(x')\| \geq (1-2\varepsilon)\|t\|.$$

Since this holds for every $\varepsilon>0$, it follows that $\|t\| = \|t_{\overline{X}'}\|$. □

From the above lemma we derive the following generalization to $\overline{\mathfrak{B}}$ of a result of Mityagin-Švarc [19], which tells us that the space $DF\overline{X}$ may always be viewed as a space of operators.

2.2. Corollary. (Mityagin-Švarc). There is an isometric inclusion

$$DF\overline{X} \to (F\overline{X}')'.$$

<u>Proof</u>: Clearly, to define this map, we follow the map $\gamma_{\overline{X}'}$ by the trace map. If $t \in DF\overline{X}$ and $x' \in F\overline{X}'$, then by 2.1,

$$\|t\| = \|t_{\overline{X}'}\| = \sup\{|\langle Tr \circ t_{\overline{X}'}, x'\rangle| \,\big|\, x' \in F\overline{X}', \|x'\| \le 1\}. \qquad \square$$

We now consider the simplest space $\overline{I}$ and the map

$$\gamma_{\overline{I}}: DF\overline{X} \to L_{L\overline{I}}(F\overline{I}, \overline{X}\otimes\overline{I}) = L(F\overline{I}, \Sigma\overline{X}).$$

To obtain the result that $\gamma_{\overline{I}}$ is mono and, hence, that $DF\overline{X}$ may be viewed as a different space of operators, we require that F be a Δ-functor, as defined by IV.4.2. Our result follows from the next proposition by taking $G=\Sigma$.

2.3. <u>Proposition</u>. (Herz-Pelletier). If F is a Δ-functor, then for every $G:\overline{\mathfrak{B}} \to \mathfrak{B}$, the map

$$NAT(F,G) \longrightarrow L(F\overline{I}, G\overline{I})$$

sending t to $t_{\overline{I}}$ is mono.

<u>Proof</u>: We consider the following diagram, which is commutative by the naturality of the evaluation map ε:

$$\begin{array}{ccc} F\overline{I}\otimes L(\overline{I},\overline{X}) & \xrightarrow{\ \varepsilon\ } & F\overline{X} \\ {\scriptstyle t_{\overline{I}}\otimes L(\overline{I},\overline{X})}\downarrow & & \downarrow{\scriptstyle t_{\overline{X}}} \\ G\overline{I}\otimes L(\overline{I},\overline{X}) & \xrightarrow{\ \varepsilon\ } & G\overline{X}. \end{array}$$

If $t_{\overline{I}} = 0$, then $t_{\overline{X}} \circ \varepsilon = 0$. Hence, if ε is epi, $t_{\overline{X}} = 0$, which proves the result. $\square$

2.4. Corollary. If F is a Δ-functor, then the map

$$\gamma_{\bar{I}}: DF\bar{X} \longrightarrow L(F\bar{I}, \Sigma\bar{X})$$

is a mono for all $\bar{X} \in \bar{\mathfrak{B}}$.

Finally, we recall from VI.2 that when $F = \mathrm{Lan}_A$, where A is an $L\bar{A}$-module, $NAT(F,G)$ is determined by the diagram $\bar{A}$, i.e.

$$NAT(\mathrm{Lan}_A, G) = L_{L\bar{A}}(A, G\bar{A}).$$

Hence, we have the following obvious result.

2.5. Proposition. $D\mathrm{Lan}_A\bar{X} = L_{L\bar{A}}(A, \bar{A}\otimes\bar{X})$ for all $\bar{X} \in \bar{\mathfrak{B}}$.

3. Duality for Computable Functors.

In this section we continue to describe $DF\bar{X}$ when certain conditions are placed on $F: \bar{\mathfrak{B}} \to \mathfrak{B}$. We shall show that if F is $\bar{\mathcal{E}}$-computable, as defined in VI.3, then $DF\bar{X}'$ can be identified as the dual of $F\bar{X}$ by means of the map

$$\Phi: DF\bar{X}' \to (F\bar{X})'$$

where $\Phi = Tr \circ \gamma_{\bar{X}}$.

In [10] Herz and Pelletier showed that when $F: \mathfrak{B} \to \mathfrak{B}$ is computable, i.e. when $FX = \varinjlim\{FY \mid Y \subseteq X, Y \in \mathcal{F}\}$, then $\Phi: DFX' \to (FX)'$ is a $\mathfrak{B}$-isomorphism. The proof of that result relied heavily on the identification of the spaces $L(E,X)'$ and $X'\otimes E$ when E is a finite dimensional Banach space. In the setting of $\overline{\mathcal{B}}$, although we have obtained the result V.4.1 stating that $\overline{X} = \varinjlim\{\overline{Y}\subseteq\overline{X} \mid \overline{Y}\in\overline{\mathcal{F}}\}$ for all $\overline{X}\in\overline{\mathcal{B}}$, making $\overline{\mathcal{F}}$-computability a natural notion, it turns out that this notion is not good for duality. We may see this by considering the obviously $\overline{\mathcal{F}}$-computable functor

$$K_0 = L(\overline{K}_0,-)$$

and seeing that duality does not hold for it since

$$(K_0\overline{X})' = L(\overline{K}_0,\overline{X})' = (\ker\sigma_0)' = X_0'/(\ker\sigma_0)^{\perp}$$

and

$$DK_0\overline{X}' = \overline{K}_0\otimes\overline{X}' = X_0'/\mathcal{C}\ell(\mathrm{im}(\sigma_0')),$$

which need not be the same unless $\mathrm{im}(\sigma_0')$ is weak* closed. This example, of course, shows that $L(-,\overline{X})'$ and $\overline{X}'\otimes-$ cannot be identified as functors when restricted to $\overline{\mathcal{F}}$, although as proved in V.4.5 they do agree on $\overline{\mathcal{E}}$. Thus, we turn our attention to $\overline{\mathcal{E}}$-computability. We notice that K_0 is not $\overline{\mathcal{E}}$-computable since for every $\overline{E}\in\overline{\mathcal{E}}$, $K_0\overline{E} = \ker\sigma_0 = 0$, implying that $\mathrm{Lan}_K UK_0 = 0$. However, $K_0 \neq 0$, since, for example, $K_0(\overline{K}_0) = I$. The following result shows that the notion of $\overline{\mathcal{E}}$-computability is the correct one for duality in the $\overline{\mathcal{B}}$-setting.

3.1. Theorem. If $F:\overline{\mathcal{B}} \to \mathcal{B}$ is $\overline{\mathcal{E}}$-computable, then

$$DF\overline{X}' = (F\overline{X})'$$

for all $\overline{X}\varepsilon\overline{\mathcal{B}}$.

Proof: Since F is $\overline{\mathcal{C}}$-computable, we have $\mathrm{Lan}_K UF = F$, where K is the inclusion $\overline{\mathcal{C}} \to \overline{\mathcal{B}}$. Because of the adjointness of Lan_K and U, we have

$$NAT(F,\overline{X}'\otimes-) = NAT(\mathrm{Lan}_K UF,\overline{X}'\otimes-) = NAT_{\overline{\mathcal{C}}}(F,\overline{X}'\otimes-).$$

Using Proposition V.4.5, as well as adjointness, we see that

$$\begin{aligned} DF\overline{X}' = NAT_{\overline{\mathcal{C}}}(F,\overline{X}'\otimes-) &= NAT_{\overline{\mathcal{C}}}(F,L(-,\overline{X})') \\ &= NAT(F,L(-,\overline{X})'). \end{aligned}$$

Since it is easy to verify that

$$NAT(F,L(-,\overline{X})') = NAT(L(-,\overline{X}),F'),$$

where $F'\overline{X} = (F\overline{X})'$, by the (contravariant) Yoneda lemma, we have

$$NAT(L(-,\overline{X}),F') = (F\overline{X})',$$

which proves our result. We leave it to the interested reader to verify that the actual isomorphism established above is the map $\Phi = Tr^{\circ}\gamma_{\overline{X}}$. □

In the classical situation where functors defined only on Banach couples are considered, the above theorem can be used to give us a duality result for the left Kan extension along the inclusion $K:\overline{\mathcal{B}\mathcal{C}} \to \overline{\mathcal{B}}$. For if $F:\overline{\mathcal{B}\mathcal{C}} \to \mathcal{B}$ is $\overline{\mathcal{C}}$-computable, then since the composition of left adjoints is again a left adjoint, $\mathrm{Lan}_K F$ will also be $\overline{\mathcal{C}}$-computable. We remark, however, that $\mathrm{Lan}_K F\overline{X} \neq F(\overline{X}/\overline{K\overline{X}})$ in general.

3.2. <u>Corollary</u>. If $F:\overline{\mathfrak{B}}\overline{\mathcal{C}} \to \mathfrak{B}$ is $\overline{\mathcal{C}}$-computable and if $\mathrm{Lan}_K F:\overline{\mathfrak{B}} \to \mathfrak{B}$ is the left Kan extension of F along the inclusion from $\overline{\mathfrak{B}}\overline{\mathcal{C}}$ to $\overline{\mathfrak{B}}$, then

$$D\mathrm{Lan}_K F\overline{X}' = (\mathrm{Lan}_K F\overline{X})'.$$

Theorem 3.1 indicates that it is useful to know when a functor is $\overline{\mathcal{C}}$-computable. We already have found in Proposition VI.5.2 a simple condition guaranteeing that Lan_A is computable. Hence, under the hypothesis of VI.5.2, we obtain from Theorem 3.1 the following result.

3.3. <u>Theorem</u>. If $\overline{A}$ is a regular Banach couple satisfying the metric approximation property, then

$$D\mathrm{Lan}_A\overline{X}' = (\mathrm{Lan}_A\overline{X})'$$

for all $\overline{X}\in\overline{\mathfrak{B}}$.

3.4. <u>Remarks</u>. (1) Theorem 3.3 is similar to a result of Janson [11]. However, since Janson considers the functor Lan^s_A (the Σ-interpolation version of Lan_A), his theorem requires an additional condition to prove, in effect, that $\mathrm{Lan}_A = \mathrm{Lan}^s_A$. (2) Theorem 3.3 also provides another proof of why duality holds for the real and complex methods of interpolation, since it will be shown in Chapter IX that these methods are actually examples of left Kan extensions, Lan_A, where A is a Banach couple satisfying the metric approximation property.

4. Approximate Reflexivity.

In Sections 2 and 3 we were able to gather information concerning $DF\overline{X}$ by putting certain conditions on F. In this section we shall introduce a condition on the doolittle diagram $\overline{X}$ which allows us to conclude that $DF\overline{X}'$ is isometrically contained in $(F\overline{X})'$ for any $F:\overline{\mathcal{B}} \to \mathcal{B}$.

4.1. Definition. We say $\overline{X}$ is approximately reflexive if for every $\overline{A}\in\overline{\mathcal{B}}$ the unit ball of $L(\overline{A},\overline{X})$ is dense in the unit ball of $L(\overline{A},\overline{X}'')$ with respect to the weak* topology given by $\overline{A}\otimes\overline{X}'$.

4.2. Remarks. (1) This definition is a generalization to $\overline{\mathcal{B}}$ of the definition of approximate reflexivity in $\mathcal{B}$ given in [10]. (2) As in the $\mathcal{B}$-setting, $\overline{X}$ is approximately reflexive if $\overline{X}$ is reflexive or if $\overline{X}'$ satisfies the metric approximation property.

4.3. Proposition. (Herz-Pelletier). $\overline{X}$ is approximately reflexive if and only if the natural map

$$\varphi:\overline{A}\otimes\overline{X}' \to L(\overline{A},\overline{X})'$$

is an isometric inclusion for every $\overline{A}\in\overline{\mathcal{B}}$.

Proof: The map $\overline{A}\otimes\overline{X}' \to L(\overline{A},\overline{X}'')'$ is an isometric inclusion since $(\overline{A}\otimes\overline{X}')' = L(\overline{A},\overline{X}'')$. Thus, the proposition amounts to the assertion that the unit ball of $L(\overline{A},\overline{X})$ is weak*-dense in the unit ball of $L(\overline{A},\overline{X}'')$. □

4.4. Proposition. (Herz-Pelletier). $\overline{X}$ is approximately reflexive if and only if for every functor $F:\overline{\mathcal{B}} \to \mathcal{B}$,

$$\Phi: DF\overline{X}' \to (F\overline{X})'$$

is an isometric inclusion.

Proof: $\Rightarrow$ Let $t \in DF\overline{X}'$. We know by 2.2 that $\|t\| = \|t_{\overline{X}''}\|$. Now choose $f \in F(\overline{X}'')$, $\|f\| \le 1$, such that $\|t_{\overline{X}''}(f)\|_{\overline{X}'\otimes\overline{X}''} \ge (1-\varepsilon)\|t_{\overline{X}''}\|$ and $T \in (\overline{X}'\otimes\overline{X}'')' = L(\overline{X}'',\overline{X}'')$ such that $\|T\| \le 1$ and

$$|\langle T, t_{\overline{X}''}(f)\rangle| = \|t_{\overline{X}''}(f)\|.$$

By approximate reflexivity, there exists $S \in L(\overline{X}'',\overline{X})$ such that $\|S\| \le 1$ and

$$|\langle T-S, t_{\overline{X}''}(f)\rangle| < \varepsilon\|t\|.$$

But then

$$|\langle S, t_{\overline{X}''}(f)\rangle| \ge |\langle T, t_{\overline{X}''}(f)\rangle| - |\langle S-T, t_{\overline{X}''}(f)\rangle| \ge (1-2\varepsilon)\|t\|.$$

Now we observe that

$$\langle S, t_{\overline{X}''}(f)\rangle = \mathrm{Tr}(t_{\overline{X}}(F(S)(f))),$$

where $F(S)(f) \in F\overline{X}$ and $\|F(S)(f)\| \le \|S\|\|f\| \le 1$. So recalling that $\Phi(t) = \mathrm{Tr}\circ t_{\overline{X}}$, we have

$$\|\Phi(t)\| \ge (1-2\varepsilon)\|t\|,$$

which proves that Φ is an isometric inclusion.

$\Leftarrow$ To prove that $\overline{X}$ is approximately reflexive, let $F = L(\overline{A},-)$, for $\overline{A}\in\mathfrak{B}$. Then by hypothesis,

$$DF\overline{X}' = \overline{A}\otimes\overline{X}' \to (F\overline{X})' = L(\overline{A},\overline{X})'$$

is an isometric inclusion. Therefore, the result follows from Proposition 4.3. □

5. Duals of Interpolation Functors.

Since the main purpose of our paper is to study interpolation functors, we shall now investigate what we can say about duals of interpolation functors. First it is essential to our theory that duality of functors preserve interpolation functors, for we wish to propose the dual functor construction as the means of obtaining the dual method of interpolation. The extent to which this is true is described in the following proposition.

5.1. Proposition. If F is a quasi-interpolation functor, then so is DF.

Proof: Since $DF(JX) = NAT(F,X\otimes\Sigma-)$, we must identify $NAT(F,X\otimes\Sigma-)$ with X. But if $x\in X$, we have a natural transformation

$$x\otimes\sigma : F \to X\otimes\Sigma-,$$

where $\sigma = F(\sigma_0,\sigma_1):F \to \Sigma$, defined by

$$(x\otimes\sigma)_{\overline{Y}}(y) = x\otimes\sigma(y).$$

The inverse of this map sends t to $t_{\overline{I}}(1) \in X\otimes I = X$. □

In view of Corollary 2.4 we have the following corollary.

5.2. <u>Corollary</u>. If F is a Δ-interpolation functor, DF is a Σ-interpolation functor.

<u>Proof</u>: By Corollary 2.4, we know that the map $DF\overline{X} \to L(F\overline{I},\Sigma\overline{X}) = \Sigma\overline{X}$ is a mono. But this map is $\sigma = DF(\sigma_0,\sigma_1)$ of the quasi-interpolation space $DF\overline{X}$. □

It is probably not true that the dual of a Σ-interpolation functor is a Δ-interpolation functor. However, we may effectively achieve that result by considering $(DF)^{\circ}$, which is by definition a Δ-interpolation functor.

It is interesting to note here that the definition of a quasi-interpolation functor F can be recast in terms of a simple demand on the values of F and DF on the single element $\overline{I}$.

5.3. <u>Proposition</u>. Let $F:\overline{\mathfrak{B}} \to \mathfrak{B}$. Then F is a quasi-interpolation functor if and only if $F\overline{I} = DF\overline{I} = I$.

<u>Proof</u>: The forward implication follows immediately from the fact that DF is a quasi-interpolation functor by 5.1.

On the other hand, if $F\overline{I} = I = DF\overline{I}$, then we obtain δ,σ directly since

$$F\overline{I} = NAT(L(\overline{I},-),F) = NAT(\Delta,F)$$

and

$$DF\overline{I} = NAT(F,\overline{I}\otimes -) = NAT(F,\Sigma).$$

Then $j = \sigma\circ\delta$ is clearly the identity on JX since $NAT(\Delta,\Sigma) = D\Delta\overline{I} = I.$ □

Now in order to combine some of the preceding results, we shall consider the interpolation functors, Lan_A, where A is a Δ-interpolation space for $\overline{A}$. This condition insures that Lan_A is a Δ-interpolation functor and, consequently, by Corollary 5.2, that $DLan_A$ is a Σ-interpolation functor. We recall from Proposition 2.5 that $DLan_A\overline{X} = L_{L\overline{A}}(A,\overline{A}\otimes\overline{X})$ for all $\overline{X}\in\mathcal{B}$.

5.4. <u>Lemma</u>. If $\overline{A}$ is non-trivial, then $L_{L\overline{A}}(\Delta\overline{A},\overline{A}\otimes\overline{X})$ is topologically isomorphic to $\Sigma\overline{X}$. If $\overline{A}$ is unital, then $L_{L\overline{A}}(\Delta\overline{A},\overline{A}\otimes\overline{X}) = \Sigma\overline{X}$.

<u>Proof</u>: By Proposition VI.2.5, $Lan_{\Delta\overline{A}} = \Delta$ when $\overline{A}$ is unital, so it follows that $DLan_{\Delta\overline{A}} = \Sigma$. The discussion in VI.2.4 tells us that $DLan_{\Delta\overline{A}}\overline{X}$ is $\mathcal{B}_\infty$-isomorphic to $\Sigma\overline{X}$ when $\overline{A}$ is non-trivial. □

We recall that in IV.5.7 the interpolation dual of the $L(\overline{A})$-module A was defined to be

$$A^* = L_{L\overline{A}}(A,\overline{A}\otimes\overline{A}'),$$

which we see is simply $DLan_A\overline{A}'$. A^* has been characterized (IV.5.9) as the maximal $L(\overline{A}')$-module contained in A'. Now we are in a position to characterize $DLan_A$ as the maximal Aronszajn-Gagliardo functor (see VI.4) with respect to A^*.

5.5. <u>Lemma</u>. If A is a Δ-interpolation space for $\bar{A}$, then

$$A^* = L_{L\bar{A}}(L\bar{A}',A') \cap \Sigma\bar{A}'.$$

<u>Proof</u>: The adjoint map takes $L_{L\bar{A}}(A,\bar{A}\otimes\bar{A}')$ to $L_{L\bar{A}}(L\bar{A}',A')$. Moreover, since $L_{L\bar{A}}(A,\bar{A}\otimes\bar{A}') = \mathrm{DLan}_A\bar{A}'$, it is contained in $\Sigma\bar{A}'$. Conversely, if $a'\in\Sigma\bar{A}'$ and if $\|Ta'\|_{A'} \le 1$ for all $T\in L\bar{A}'$, $\|T\| \le 1$, then

$$|\langle T,a\otimes a'\rangle| \le \|a\|_A \quad \text{for all} \quad a\in A,$$

so $\|a'\|_{L_{L\bar{A}}(A,\bar{A}\otimes\bar{A}')} \le 1$. □

5.6. <u>Theorem</u>. If $\bar{A}$ is non-trivial and if A is a Δ-interpolation space for $\bar{A}$, then $L_{L\bar{A}}(A,\bar{A}\otimes\bar{X})$ is topologically isomorphic to $L_{L\bar{A}'}(L(\bar{X},\bar{A}'),A^*) \cap \Sigma\bar{X}$. If $\bar{A}$ is unital, then

$$L_{L\bar{A}}(A,\bar{A}\otimes\bar{X}) = L_{L\bar{A}'}(L(\bar{X},\bar{A}'),A^*) \cap \Sigma\bar{X}, \text{ i.e.}$$

$$\mathrm{DLan}_A\bar{X} = H_{A^*}\bar{X}.$$

<u>Proof</u>: We shall prove the second statement. We have $L_{L\bar{A}}(A,\bar{A}\otimes\bar{X}) = \mathrm{DLan}_A\bar{X} \subset \Sigma\bar{X}$. In fact $L_{L\bar{A}}(A,\bar{A}\otimes\bar{X})$ may be interpreted as the set

$$\{x\in\Sigma\bar{X} \mid \|a\otimes x\|_{\bar{A}\otimes\bar{X}} \le C\|a\|_A\}.$$

We can also define

$$\psi : L_{L\overline{A}}(A,\overline{A}\otimes\overline{X}) \to L_{L\overline{A}}(L(\overline{X},\overline{A}'),A')$$

by

$$\langle\psi(x)(T),a\rangle = \langle Tx,a\rangle.$$

It follows that if $\|x\|_{L_{L\overline{A}}(A,\overline{A}\otimes\overline{X})} \le 1$ and if $\|T\| \le 1$, then $\langle Tx,a\rangle = \langle T,a\otimes x\rangle \le \|a\|_A$, i.e. $\|Tx\| \le 1$. However, if $S\in L\overline{A}'$, $\|S\| \le 1$, then

$$\psi(x)(S\circ T) = S\psi(x)(T)$$

and $\|STx\| \le \|Tx\|_{A'}$, so $Tx \in L_{L\overline{A}}(L\overline{A}',A') = A^*$. Therefore, we conclude that ψ actually takes $L_{L\overline{A}}(A,\overline{A}\otimes\overline{X})$ to $L_{L\overline{A}'}(L(\overline{X},\overline{A}'),A^*)$. The proof of the converse follows the pattern of Lemma 5.5. □

We recall that A^* was originally defined in order to have a reasonable dual interpolation space for given A since in general A' might fail to be an $L\overline{A}'$-module. When A^* is actually equal to A', we have the following result.

5.7. <u>Theorem</u>. If $A^* = A'$ and if $\overline{X}$ is approximately reflexive, then

$$(\mathrm{Lan}_A\overline{X})' = D\mathrm{Lan}_A\overline{X}'.$$

<u>Proof</u>: We know by Proposition 4.4 that $\Phi : D\mathrm{Lan}_A\overline{X}' \to (\mathrm{Lan}_A\overline{X})'$ is an isometric inclusion, and, of course, that $(\mathrm{Lan}_A\overline{X})' \subset \Sigma\overline{X}'$. Then letting $x' \in (\mathrm{Lan}_A\overline{X})'$, we must show that x' defines an element of $D\mathrm{Lan}_A\overline{X}' = L_{L\overline{A}}(A,\overline{A}\otimes\overline{X}')$. It suffices to show that if $a_0 \in \Delta\overline{A}$, then $\|a_0\otimes x'\|_{\overline{A}\otimes\overline{X}'} \le C\|a_0\|_A$ for some C. Let $\|a_0\|_{\Delta\overline{A}} \le 1$. We have

$$\|a_0\otimes x'\|_{\overline{A}\otimes\overline{X}'} = \sup\{|\langle T,a_0\otimes x'\rangle| \,\big|\, T\in L(\overline{A},\overline{X}''),\ \|T\|\le 1\},$$

and since the unit ball of $L(\overline{A},\overline{X})$ is dense in the unit ball of $L(\overline{A},\overline{X}'')$, this is equal to

$$\begin{aligned}&\sup\{|\langle T,a_0\otimes x'\rangle| \,\big|\, T\in L(\overline{A},\overline{X}),\ \|T\|\le 1\}\\ \le\ &\sup\{|\langle x',Ta\rangle| \,\big|\, T\in L(\overline{A},\overline{X}),\ a\in A,\ \|T\|\|a\|\le 1\}\\ &= \|x'\|_{(\mathrm{Lan}_A\overline{X})'}.\end{aligned}$$

Therefore, $\|x'\|_{\mathrm{DLan}_A\overline{X}'} \le \|x'\|_{(\mathrm{Lan}_A\overline{X})'}$. □

Finally, we close by pointing out a consequence of our work which gives one condition under which $A' = A^*$, i.e. under which A' is an $L\overline{A}'$-module. This result also can be deduced from Janson's duality theorem as well as our Theorem 3.3, although he never states this result explicitly. Of course, under the hypothesis of 3.3, Theorem 5.7 tells us nothing that does not already follow from the former theorem.

5.8. <u>Proposition</u>. If $\overline{A}$ is a regular Banach couple satisfying the metric approximation property and if A is a Δ-interpolation space for $\overline{A}$, then $\mathrm{DLan}_A\overline{A}' = A'$. Hence, in particular, A' is actually an interpolation space for $\overline{A}'$ and, thus, $A' = A^*$.

PART III

CHAPTER VIII

MORE ABOUT DUALITY

1. Comparison of Parts I and II.

To this point in our work we have studied extensions of functors and duality questions twice. In Chapters II and III we were interested in extending the classical interpolation methods to the $\overline{\mathfrak{B}}$-setting and establishing that the classical duality results remained valid. With respect to the complex method of interpolation, we extended the classical C_θ-, C^θ-methods to $\overline{\mathfrak{B}}$ naively using the following fact: if F is an interpolation functor defined on $\overline{\mathfrak{B}}\mathfrak{C}$, then $\tilde{F}$ defined by $\tilde{F}(\overline{X}) = F(\overline{X}/\overline{K}\overline{X})$ is an interpolation functor on $\overline{\mathfrak{B}}$. We then proved a General Duality Theorem (III.1.3) which gave us conditions under which the classical duality results held in the larger setting. On the other hand, since most interpolation functors arising in practice are either minimal or maximal, the method of extending functors by the Kan extension construction, examined in Chapter VI, is more faithful to reality, albeit more sophisticated. Our abstract study of duality undertaken in Chapter VII established various duality results for the functors Lan_A and H_A, which are the categorical versions of the minimal and maximal Aronszajn-Gagliardo functors, respectively, arising from Kan extensions. In particular, we proved that the dual functor of Lan_A is equal to H_{A^*} (VII.5.6), where A^* is the interpolation dual $L_{L\overline{A}}(A,\overline{A}\otimes\overline{A}')$, and that

$DLan_A\overline{X}' = (Lan_A\overline{X})'$ whenever $\overline{A}$ is a regular Banach couple satisfying the metric approximation property (VII.3.3) or whenever $A' = A^*$ and $\overline{X}$ is approximately reflexive (VII.5.7).

Thus it is natural and interesting for us to find the relations between the results of Parts I and II. We have two particular questions in mind. The first is to know when the two conceptually different approaches to the extension of an interpolation functor on $\overline{\mathfrak{BC}}$ lead to the same results. The second is to know when the General Duality Theorem is applicable to the functors $F = Lan_A$ and $G = DLan_A$. Both questions require finding out when $Lan_A\overline{X} = Lan_A(\overline{X}/\overline{K}\overline{X})$ and $DLan_A\overline{X} = DLan_A(\overline{X}/\overline{K}\overline{X})$. Moreover, the second question involves finding the conditions under which $Lan_A\overline{X} = Lan_A\overline{X}^0$ and $DLan_A\overline{X} = DLan_A\overline{X}^0$, which reinforces our general viewpoint that the connection between $\overline{X}$ and $\overline{X}^0$ is as important as that between $\overline{X}$ and $\overline{X}/\overline{K}\overline{X}$.

It turns out that two of these equalities are immediate consequences of the regularity of $\overline{A}$ and involve no special relation between $\overline{A}$ and the $L\overline{A}$-module A.

1.1. <u>Proposition</u>. If $\overline{A}=\overline{A}^0$, then $L(\overline{A},\overline{X}) = L(\overline{A},\overline{X}^0)$ and $\overline{A}\otimes\overline{X} = \overline{A}\otimes(\overline{X}/\overline{K}\overline{X})$ so $Lan_A\overline{X} = Lan_A\overline{X}^0$ and $DLan_A\overline{X} = DLan_A(\overline{X}/\overline{K}\overline{X})$.

<u>Proof</u>: Let $T = (T_0,T_1) \in L(\overline{A},\overline{X})$. We have to prove that $T \in L(\overline{A},\overline{X}^0)$, i.e. that $T_i \in L(A_i,X_i^0)$, i=0,1.

Let $x_i \in X_i$ and suppose that $x_i = T_i a_i$ for some $a_i \in A_i$. By assumption there exists a sequence $\{d_k\} \subset \Delta\overline{A}$ such that $\|\delta_i d_k - a_i\|_{A_i} \to 0$. But then $T_i(\delta_i d_k) = \delta_i(Td_k) \in X_i$, and since

$$\|T_i a_i - T_i(\delta_i d_k)\| = \|T_i(a_i - \delta_i d_k)\| \le \|T_i\|\|a_i - \delta_i d_k\| \to 0,$$

it follows that $x_i \in \mathcal{C}\ell(\mathrm{im}(\delta_i)) = X_i^0$. Hence, $T \in L(\overline{A},\overline{X}^0)$.

To see that $\overline{A}\otimes\overline{X} = \overline{A}\otimes(\overline{X}/\overline{KX})$, it suffices to observe that

$$(\overline{A}\otimes\overline{X})' = L(\overline{A},\overline{X}') = L(\overline{A},(\overline{X}')^0) = L(\overline{A},((\overline{X}/\overline{KX})')^0) = (\overline{A}\otimes(\overline{X}/\overline{KX}))'. \quad \square$$

The question of determining when $\mathrm{Lan}_A\overline{X} = \mathrm{Lan}_A(\overline{X}/\overline{KX})$ and $\mathrm{DLan}_A\overline{X} = \mathrm{DLan}_A\overline{X}^0$ is much deeper, and we suspect that the sufficient conditions which we find to insure the results are also very close to being necessary. If this is indeed the case, then a pair (F,G) of interpolation functors satisfying the General Duality Theorem with $F = \mathrm{Lan}_A$ is actually not very different from the real or complex methods of interpolation, although we shall not make this conjecture precise.

We shall start by making some progress towards determining when $\mathrm{DLan}_A\overline{X} = \mathrm{DLan}_A\overline{X}^0$.

1.2. <u>Lemma</u>. Let $\overline{A}$ be a regular Banach couple. Suppose that $\Delta\overline{A}$ is dense in A and that $A \not\subset A_0$. Then for all $\overline{X}\in\mathfrak{B}$, we have

$$\mathrm{DLan}_A\overline{X} \subset \sigma_0(X_0^0) + \sigma_1(X_1) \subset \Sigma\overline{X}.$$

<u>Proof</u>: Because of the density of $\Delta\overline{A}$ in A, the condition that A is not contained in A_0 means that the map $\delta_0:\Delta\overline{A} \to A_0$ is not continuous when $\Delta\overline{A}$ is given the norm from A. Hence, there exists for each n an element $a_n \in \Delta\overline{A}$ such that $\|\delta a_n\|_A = 1$ and $\|\delta_0 a_n\| = n$. By the Hahn-Banach theorem, there exists $a_n' \in A_0'$ such that $\|a_n'\| = 1$ and $\langle a_n',\delta_0 a_n\rangle = n$.

Now suppose that we have $x \in \mathrm{DLan}_A\overline{X} = L_{L\overline{A}}(A,\overline{A}\otimes\overline{X})$ such that $x \notin \sigma_0(X_0^0) + \sigma_1(X_1)$, i.e. for every representation $x = x_0 + x_1$, we have $x_0 \notin X_0^0$. Then given such a representation of x, there exists $x_0' \in (\mathrm{im}(\delta_0))^\perp \subset X_0'$ such that $\|x_0'\| = 1$ and $|\langle x_0',x_0\rangle| \neq 0$. But since $x \in L_{L\overline{A}}(A,\overline{A}\otimes\overline{X})$, we have

$$\|a\otimes x\|_{\overline{A}\otimes\overline{X}} = \sup\{|\langle Tx,a\rangle| \mid T \in L(\overline{X},\overline{A}'),\ \|T\| \le 1\}$$

$$(*) \qquad \le \|x\|_{DLan_A\overline{X}}\|a\|_A.$$

However, for every $a_0' \in A_0'$ such that $\|a_0'\| = 1$, we have $x_0'\otimes a_0' \in L(\overline{X},\overline{A}')$ and $\|x_0'\otimes a_0'\|_{L(\overline{X},\overline{A}')} = \|x_0'\|\|a_0'\|$. Therefore, choosing $a = a_n$ and $a_0' = a_n'$, we have

$$\langle (x_0'\otimes a_n')x, a_n\rangle = \langle x_0',x\rangle\langle a_n',a_n\rangle = n\langle x_0',x\rangle \to \infty,$$

which contradicts (*). □

1.3. Corollary. If $\overline{A}$ is a regular Banach couple with $\Delta\overline{A}$ dense in A, then $DLan_A\overline{X} \subset \Sigma\overline{X}^0$ for all $\overline{X}$ if and only if $A \not\subset A_0 \cup A_1$.

2. Quasi-injectivity and Quasi-projectivity.

The condition that arises in obtaining the equality of $DLan_A\overline{X}$ and $DLan_A\overline{X}^0$ is that of injectivity, while the dual notion of projectivity arises with respect to the equality of $Lan_A\overline{X}$ and $Lan_A(\overline{X}/\overline{KX})$.

2.1. Definitions. (1) $\overline{A}$ is said to be quasi-injective (compactly quasi-injective) if A_i, i=0,1, is (compactly) injective in $\mathcal{B}$ with respect to isometric inclusions, i.e. if whenever $X \to Y$ is an isometric inclusion, any (compact) map $T: X \to A_i$ can be extended to a (compact) map $\tilde{T}: Y \to A_i$ with $\|\tilde{T}\| = \|T\|$ ($\|\tilde{T}\| < (1+\varepsilon)\|T\|$). (2). $\overline{A}$ is said to be quasi-projective (compactly quasi-projective) if A_i,

$i=0,1$, is (compactly) projective with respect to quotient maps, i.e. if whenever $X \to Y$ is a quotient map, any (compact) map $S:A_i \to Y$ can be lifted to a (compact) map $\tilde{S}:A_i \to X$ with $\|\tilde{S}\| < (1+\varepsilon)\|S\|$.

We state without proof the following straightforward proposition.

2.2. <u>Proposition</u>. If $\bar{A}$ is (compactly) quasi-projective, then $\bar{A}'$ is (compactly) quasi-injective. Moreover, if $\bar{A}$ is compactly quasi-projective and if $\bar{A}'$ satisfies the metric approximation property, then $\bar{A}'$ is quasi-injective.

Letting $\mathbb{K}(\bar{A},\bar{X})$ denote the compact continuous linear maps T from $\bar{A}$ to $\bar{X}$ (i.e. T_0,T_1 are compact), then if $q:\bar{X} \to \bar{X}/\overline{KX}$ is the natural quotient map, we obtain maps

$$q^* = L(\bar{A},q):L(\bar{A},\bar{X}) \to L(\bar{A},\bar{X}/\overline{KX})$$

and

$$q^* = \mathbb{K}(\bar{A},q):\mathbb{K}(\bar{A},\bar{X}) \to \mathbb{K}(\bar{A},\bar{X}/\overline{KX}).$$

2.3. <u>Proposition</u>. If $\bar{A}$ is (compactly) quasi-projective, then $q^*:L(\bar{A},\bar{X}) \to L(\bar{A},\bar{X}/\overline{KX})$ $(q^*:\mathbb{K}(\bar{A},\bar{X}) \to \mathbb{K}(\bar{A},\bar{X}/\overline{KX}))$ is a quotient map.

<u>Proof</u>: Given $T = (T_0,T_1) \in L(\bar{A},\bar{X}/\overline{KX})$, we can by the projectivity of A_i lift T_i to $\tilde{T}_i:A_i \to X_i$ with $\|\tilde{T}_i\| < (1+\varepsilon)\|T\|$. Then we define $\tilde{T} = (\tilde{T}_0,\tilde{T}_1)$, and we observe that since $\sigma_i:X_i \to \Sigma\bar{X}$ factors through $X_i/K_i\bar{X}$, it follows that $\sigma_0\circ\tilde{T}_0\circ\delta_0 = \sigma_1\circ\tilde{T}_1\circ\delta_1$, so $\tilde{T} \in L(\bar{A},\bar{X})$. The analogous argument works for compact maps as well. □

2.4. <u>Corollary</u>. If $\bar{A}$ is quasi-projective (or if $\bar{A}$ is compactly quasi-projective and if $\mathrm{Lan}_A\bar{X} = \mathbb{K}(\bar{A},\bar{X})\otimes_{L\bar{A}}A$), then

$$q^{*}\otimes 1_A : Lan_A\overline{X} \to Lan_A(\overline{X}/\overline{KX})$$

is a quotient map.

We are now in a position to prove our result on the equality of $DLan_A\overline{X}$ and $DLan_A\overline{X}^0$.

2.5. <u>Proposition</u>. If $\overline{A}$ is a regular Banach couple with $\Delta\overline{A}$ dense in A, $A \not\subset A_0 \cup A_1$, and with $\overline{A}'$ quasi-injective, then $DLan_A\overline{X} = DLan_A\overline{X}^0$.

<u>Proof</u>: We recall that by 1.3, $DLan_A\overline{X} \subset \Sigma\overline{X}^0$. Hence, we must show that the map

$$DLan_A\overline{X}^0 \to DLan_A\overline{X}$$

is onto, i.e. that $x \in DLan_A\overline{X} \subset \Sigma\overline{X}^0$ actually defines an element of $DLan_A\overline{X}^0$. We have for any $a \in \Delta\overline{A}$,

$$\|a\otimes x\|_{\overline{A}\otimes\overline{X}^0} = \sup\{|\langle Tx,a\rangle| \,|\, T \in L(\overline{X}^0,\overline{A}'),\ \|T\| \le 1\}.$$

But by the quasi-injectivity of $\overline{A}'$, we may extend $T_i : X_i^0 \to A_i'$ to $T_i : X_i \to A_i'$, to get $T \in (\overline{X},\overline{A}')$, $\|T\| \le 1$. Therefore,

$$\begin{aligned}\|a\otimes x\|_{\overline{A}\otimes\overline{X}^0} &= \sup\{|\langle Tx,a\rangle| \,|\, T \in L(\overline{X},\overline{A}'),\ \|T\| \le 1\} \\ &= \|a\otimes x\|_{\overline{A}\otimes\overline{X}}.\end{aligned}$$ □

Finally, to get the equality of $Lan_A\overline{X}$ and $Lan_A(\overline{X}/\overline{KX})$, we need a condition which involves the full structure of $\overline{A}$ and A.

2.6. <u>Theorem</u>. Suppose that (i) $\overline{A}$ is quasi-projective or (ii) $\overline{A}$ is compactly quasi-projective and that $Lan_A\overline{X} = \mathbb{K}(\overline{A},\overline{X})\otimes_{L\overline{A}}A$. Assume further that there exist $U, V \in L\bar{A}$ such that U, V are both invertible with $\|U\|_{A_0}\|U^{-1}\|_A < 1$ and $\|V\|_{A_1}\|V^{-1}\|_A < 1$. Then

$$\mathrm{Lan}_A\bar{X} = \mathrm{Lan}_A(\bar{X}/\bar{K}\bar{X}).$$

<u>Proof</u>: We know from 2.4 that $q^*\otimes 1_A : \mathrm{Lan}_A\bar{X} \to \mathrm{Lan}_A(\bar{X}/\bar{K}\bar{X})$ is a quotient. In order to prove that $q^*\otimes 1_A$ is also injective, we shall consider the dual map

$$(q^*\otimes 1_A)' : (\mathrm{Lan}_A(\bar{X}/\bar{K}\bar{X}))' = L_{L\bar{A}}(A, L(\bar{A}, \bar{X}/\bar{K}\bar{X})') \to (\mathrm{Lan}_A\bar{X})' = L_{L\bar{A}}(A, L(\bar{A}, \bar{X})')$$

and show that it is an isomorphism. We first observe that since Lan_A is a Δ-interpolation functor, $(\mathrm{Lan}_A\bar{X})'$ and $(\mathrm{Lan}_A(\bar{X}/\bar{K}\bar{X}))'$ are contained, respectively, in $\Sigma\bar{X}'$ and $\Sigma(\bar{X}/\bar{K}\bar{X})' = \Sigma\bar{K}^{\perp} \subset \Sigma\bar{X}'$. In fact, since $\Sigma\bar{X}' = (\mathrm{Lan}_{\Delta\bar{A}}\bar{X})'$, we see that $f \in \Sigma\bar{X}' = L_{L\bar{A}}(\Delta\bar{A}, L(\bar{A}, \bar{X})')$ will be in $(\mathrm{Lan}_A\bar{X})'$ if $\|fa\|_{L(\bar{A},\bar{X})'} \le C\|a\|_A$ for all $a \in \Delta\bar{A}$, i.e. if $|\langle fa, T\rangle| \le C\|a\|_A\|T\|$ for all $a \in \Delta\bar{A}$, $T \in L(\bar{A}, \bar{X})$. However, $\langle fa, T\rangle = \langle f, Ta\rangle$, for $f \in \Sigma\bar{X}'$, $a \in \Delta\bar{A}$, $T \in L(\bar{A}, \bar{X})$, so $f \in (\mathrm{Lan}_A\bar{X})'$ if

$$\|f\|_{(\mathrm{Lan}_A\bar{X})'} = \sup\{|\langle f, Ta\rangle| \mid a \in \Delta\bar{A},\ \|a\|_A \le 1,\ T \in L(\bar{A}, \bar{X}),\ \|T\| \le 1\} < \infty.$$

By similar arguments, $g \in (\mathrm{Lan}_A(\bar{X}/\bar{K}\bar{X}))'$ if $g \in \Sigma\bar{K}^{\perp}$ and if

$$\|g\|_{(\mathrm{Lan}_A(\bar{X}/\bar{K}\bar{X}))'} = \sup\{|\langle g, Ta\rangle| \mid a \in \Delta\bar{A},\ \|a\|_A \le 1,\ T \in L(\bar{A}, \bar{X}/\bar{K}\bar{X}),\ \|T\| \le 1\} < \infty.$$

To complete the proof we thus have to prove that if $f \in (\mathrm{Lan}_A\bar{X})'$, then $f \in \Sigma\bar{K}^{\perp}$ and that $\|f\|_{(\mathrm{Lan}_A\bar{X})'} = \|f\|_{(\mathrm{Lan}_A\bar{X}/\bar{K}\bar{X})'}$. For this it suffices to prove that if $T \in L(\bar{A}, \bar{X})$, $a \in \Delta\bar{A}$, then

$$\|T\otimes a\|_{\mathrm{Lan}_A\overline{X}} \le \|q^*T\|\|a\|_A.$$

This is where the assumptions of the proposition are used. For if $\overline{A}$ is quasi-projective, there exists $S \in L(\overline{A},\overline{X})$ such that $q^*(S) = q^*(T)$ and $\|S\| < (1+\varepsilon)\|q^*(T)\|$. (If $\overline{A}$ is compactly quasi-projective, then we may assume that T is compact. Then a compact S exists with the above property.) But then $q^*(T-S) = 0$, so $T-S \in L(\overline{A},\overline{K}\overline{X})$. Let us write $T-S = R_0 + R_1$, where $R_0 \in L(A_0,K_0\overline{X})$ and $R_1 \in L(A_1,K_1\overline{X})$. Then

$$T\otimes a = S\otimes a + (T-S)\otimes a = S\otimes a + R_0\otimes a + R_1\otimes a.$$

Obviously, $\|S\otimes a\|_{\mathrm{Lan}_A\overline{X}} \le \|S\|\|a\|_A < (1+\varepsilon)\|q^*T\|\|a\|_A$, so it suffices to prove that

$$\|R_0\otimes a\|_{\mathrm{Lan}_A\overline{X}} = 0 = \|R_1\otimes a\|_{\mathrm{Lan}_A\overline{X}}.$$

Since the arguments are the same, we shall prove that $\|R_0\otimes a\| = 0$. Using the element $U \in L\overline{A}$ of the hypothesis, we have

$$R_0\otimes a = R_0\otimes U^nU^{-n}a = R_0\circ U^n\otimes U^{-n}a,$$

so

$$\|R_0\otimes a\| \le \|R_0U^n\|\|U^{-n}a\|_A \le \|R_0\|\|U\|^n_{A_0}\|U^{-1}\|^n_A\|a\| < \varepsilon$$

if n is sufficiently large.

This proves the proposition. □

In closing this chapter we wish to remark that since the General Duality Theorem involves no approximation property conditions on $\overline{X}$, for pairs $(\overline{A}, A)$ satisfying the various assumptions above (1.1, 2.5, 2.6), we obtain a duality result for Lan_A, $DLan_A$ which is free from any approximation restrictions.

CHAPTER IX

THE CLASSICAL METHODS FROM A CATEGORICAL VIEWPOINT

1. Review of Results.

In Part I of this paper we showed that the classical real and complex methods have natural extensions from the category of Banach couples to the category of doolittle diagrams of Banach spaces. We also showed that the equivalence for the real method still holds and that the classical duality theorems are true. Moreover, it is easy to prove that all the classical theorems retain their validity.

In the second part of the paper we made a detailed study of the category $\overline{\mathfrak{B}}$ and demonstrated that it was a natural setting in which to define interpolation functors. The notion of dual functor available in $\overline{\mathfrak{B}}$ was an important tool in realizing our aim of finding a setting of interpolation theory in which duality plays a real role. In order to have a good duality theory, we defined two classes of interpolation functors, the Δ-interpolation and Σ-interpolation functors, noting that most of the classical functors are actually in both classes. The dual functor turned out to be well adapted for Δ-interpolation functors, while it required a modification in the case of Σ-interpolation functors.

We also investigated in Part II the idea of extending a functor defined only on a subcategory of $\overline{\mathfrak{B}}$ in the categorical sense of forming its left or right Kan extension. These extensions are minimal and maximal, respectively, and are related to the classical notion, due to Aronszajn and Gagliardo, of minimal and maximal extensions of interpolation functors. However, we found that neither the Kan extensions nor the Aronszajn-Gagliardo functors were correct for our theory and we instead used a mixture of them, namely the left Kan extension and the maximal Aronszajn-Gagliardo functor. Our main abstract result (VII.5.6) was that if a Δ-interpolation functor is a left Kan extension, then its dual functor is a maximal Aronszajn-Gagliardo functor.

A very important result in abstract interpolation theory was the discovery (due to Brudnyi-Krugljak [3] and independently to Janson [11]) that the real and complex methods are Aronszajn-Gagliardo functors. We shall investigate here to what extent they are left Kan extensions or maximal Aronszajn-Gagliardo functors and whether the J- and K- and C_θ- and C^θ- methods are dual functors of each other.

2. The Real Method Revisited.

In Chapter II we proved that the classical duality theorem for the J- and K-methods held in our setting. However, we did not prove that these two methods are dual functors. We shall in this section be able to prove that this is the case by means of the fact that the J-method is actually a left Kan extension from a particular doolittle diagram (actually a regular Banach couple). We are going to consider only the discrete real methods.

In order to define the appropriate doolittle diagram for obtaining the J-method, we begin by introducing some useful spaces.

2.1. <u>Definitions</u>. (1) For $1 \leq q < \infty$ and $\theta \in \mathbb{R}$, we let λ_θ^q denote the space

$$\lambda_\theta^q = \{\{x_k\}_{k\varepsilon\mathbf{Z}} \mid (\sum_{-\infty}^{\infty} (2^{\theta k}|x_k|)^q)^{1/q} = \|x\|_{q,\theta} < \infty\},$$

and

$$\lambda_\theta^\infty = \{\{x_k\}_{k\varepsilon\mathbf{Z}} \mid \sup(2^{\theta k}|x_k|) = \|x\|_{\infty,\theta} < \infty\}.$$

(2) We write λ_c to denote the space

$$\lambda_c = \{\{x_k\}_{k\varepsilon\mathbf{Z}} \mid x_k = 0 \text{ for all but finitely many } k\text{'s}\}.$$

We observe that λ_c is dense in λ_θ^q unless $q=\infty$. The standard basis vectors in λ_θ^q will be denoted by e_k. For convenience we shall define a duality between λ_θ^q and $\lambda_\theta^{q'}$ (where $1/q + 1/q' = 1$) by

$$\langle\{x_k\}, \{y_k\}\rangle = \sum_{-\infty}^{\infty} x_k y_{-k},$$

for $\{x_k\} \in \lambda_\theta^q$ and $\{y_k\} \in \lambda_\theta^{q'}$.

2.2. <u>Definition</u>. The doolittle diagram $\bar{\lambda}^1$ is defined as follows: we first form the pushout $\Sigma\bar{\lambda}^1$ of the diagram

$$\begin{array}{ccc} \lambda_c & \longrightarrow & \lambda_0^1 \\ \downarrow & & \\ \lambda_1^1 & & \end{array}$$

and then take the pullback $\Delta\bar{\lambda}^1$ of the diagram

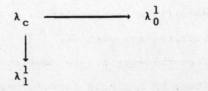

$$\begin{array}{ccc} & & \lambda_0^1 \\ & & \downarrow \\ \lambda_1^1 & \longrightarrow & \Sigma\bar{\lambda}^1. \end{array}$$

2.3. Remarks. (1) It is easy to see by definition of the pushout and pullback that

$$\Sigma\bar{\lambda}^1 = \{\{x_k\} \mid \sum_{-\infty}^{\infty} \min(1,2^k)|x_k| = \|x\| < \infty\}$$

and

$$\Delta\bar{\lambda}^1 = \{\{x_k\} \mid \max(\sum_{-\infty}^{\infty} |x_k|, \sum_{-\infty}^{\infty} 2^k|x_k|) < \infty\}.$$

We note that the norm on $\Delta\bar{\lambda}^1$ is equivalent to the norm $\sum_{-\infty}^{\infty} \max(1,2^k)|x_k|$. (2) Clearly, $\bar{\lambda}^1$ is a regular Banach couple.

To see why $\bar{\lambda}^1$ is related to the real method we shall first calculate $L(\bar{\lambda}^1,\bar{X})$ and $\bar{\lambda}^1\otimes\bar{X}$.

2.4. Proposition. Every $T \in L(\bar{\lambda}^1,\bar{X})$ is determined by a sequence $\{x_k\}_{k\in\mathbb{Z}}$ in $\Delta\bar{X}$ such that $\sup J(2^{-k},x_k) < \infty$.

Proof: Let $T = (T_0,T_1) \in L(\bar{\lambda}^1,\bar{X})$. Since λ_i^1 is an ℓ^1-space, $i=0,1$, $T_i \in L(\lambda_i^1,X_i)$ is given by a sequence $\{x_{ik}\}_{k\in\mathbb{Z}}$, where $x_{ik} = T_i(e_k)$. Moreover, $\|T_0\| = \sup\|x_{0k}\|_{X_0}$ and $\|T_1\| = \sup 2^{-k}\|x_{1k}\|_{X_1}$ since $\|e_k\|_{\lambda_1^1} = 2^k$. Furthermore, since $\sigma_0T_0(e_k) = \sigma_0(x_{0k}) = \sigma_1T_1(e_k) = \sigma_1(x_{1k})$, there exists $x_k\in\Delta\bar{X}$ such that $\delta_i(x_k) = x_{ik}$, $i=0,1$. Clearly, the sequence $\{x_k\}$ describes the behaviour of T and also

$$\begin{aligned}\sup J(2^{-k},x_k) &= \sup \max(\|\delta_0 x_k\|,\ 2^{-k}\|\delta_1 x_k\|)\\ &= \max \sup(\|\delta_0 x_k\|,\ 2^{-k}\|\delta_1 x_k\|)\\ &= \|T\| < \infty.\end{aligned}$$

Conversely, if $\{x_k\}_{k\varepsilon\mathbf{Z}}$ is a sequence in $\Delta\bar{X}$ such that $\sup J(2^{-k},x_k) < \infty$, then we can define $T \in L(\bar{\lambda}^1,\bar{X})$ by $T(e_k) = x_k$. □

2.5. <u>Proposition</u>. The space $\bar{\lambda}^1\otimes\bar{X}$ consists of all sequences $\{x_k\}_{k\varepsilon\mathbf{Z}}$ in $\Sigma\bar{X}$ such that $\Sigma K(2^k,x_k) < \infty$.

<u>Proof</u>: Since the space $\lambda_i^1\otimes X_i$ is an ℓ^1-space, we can write

$$\lambda_0^1\otimes X_0 = \{\{x_{0k}\} \mid \sum_{-\infty}^{\infty} \|x_{0k}\|_{X_0} < \infty\}$$

and

$$\lambda_1^1\otimes X_1 = \{\{x_{1k}\} \mid \sum_{-\infty}^{\infty} 2^k\|x_{1k}\|_{X_1} < \infty\}.$$

Also, the space $\Delta\bar{\lambda}^1\otimes\Delta\bar{X}$ is the space

$$\Delta\bar{\lambda}^1\otimes\Delta\bar{X} = \{\{x_k\} \mid x_k\in \Delta\bar{X},\ \sum_{-\infty}^{\infty} \max(1,2^k)\|x_k\| < \infty\}.$$

Since $\bar{\lambda}^1\otimes\bar{X}$ is the pushout of the diagram

$$\begin{array}{ccc}\Delta\bar{\lambda}^1\otimes\Delta\bar{X} & \longrightarrow & \lambda_0^1\otimes X_0\\ \downarrow & & \\ \lambda_1^1\otimes X_1 & & \end{array},$$

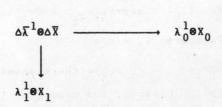

we obtain that

$$\bar{\lambda}^{1}\otimes\bar{X} = \{\{x_k\} \mid x_k \in \Sigma\bar{X},\ \Sigma K(2^{k},x_k) < \infty\}. \qquad \square$$

We note here that the dual of the diagram $\bar{\lambda}^{1}$ is the diagram

$$\bar{\lambda}^{\infty} = \begin{array}{ccc} \Delta\bar{\lambda}^{\infty} & \longrightarrow & \lambda_0^{\infty} \\ \Big\downarrow & & \Big\downarrow \\ \lambda_1^{\infty} & \longrightarrow & \Sigma\bar{\lambda}^{\infty} \end{array},$$

where $\Delta\bar{\lambda}^{\infty} = (\Sigma\bar{\lambda}^{1})'$ and $\Sigma\bar{\lambda}^{\infty} = (\Delta\bar{\lambda}^{1})'$.

Now we shall apply the $J(\theta,q)$-method to $\bar{\lambda}^{1}$. According to the classical definition, this is the subspace of $\Sigma\bar{\lambda}^{1}$ consisting of all sequences $x = \{x_k\}_{k\in\mathbf{Z}}$ such that $x = \sum_{-\infty}^{\infty} u_m$, where $u_m \in \Delta\bar{\lambda}^{1}$ and $\left(\sum_{-\infty}^{\infty} (2^{\theta m}J(2^{-m},u_m))^{q}\right)^{1/q} < \infty$. If $x = \{x_k\} \in \lambda_{\theta}^{q}$, then we can write $x = \sum_{-\infty}^{\infty} x_k e_k = \sum_{-\infty}^{\infty} u_k$ and since $J(2^{-k},u_k) = |x_k|$, we have

$$\left(\sum_{-\infty}^{\infty} (2^{\theta k}J(2^{-k},u_k))^{q}\right)^{1/q} = \left(\sum_{-\infty}^{\infty} (2^{\theta k}|x_k|)^{q}\right)^{1/q} = \|x\|_{q,\theta}.$$

This shows that $\lambda_{\theta}^{q} \subset J(\theta,q,\bar{\lambda}^{1})$ and that

$$\|x\|_{J(\theta,q,\bar{\lambda}^{1})} \le \|x\|_{\theta,q}.$$

We will denote the map from λ_{θ}^{q} to $J(\theta,q,\bar{\lambda}^{1})$ by $\alpha_{\theta,q}$.

For $q=1$ the situation is more precise.

2.6. Proposition. $J(\theta,1,\bar{\lambda}^{1}) = \lambda_{\theta}^{1}$.

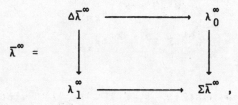

<u>Proof</u>: Let $x \in J(\theta,1,\bar{\lambda}^1)$. Then by the description above there exists $u_k \in \Delta\bar{\lambda}^1$ such that $x = \sum_{-\infty}^{\infty} u_k$ and $\sum_{-\infty}^{\infty} 2^{\theta k} J(2^{-k},u_k) < \infty$. Let us write $u_k = \{x_{km}\}_{m\in\mathbb{Z}}$. Then

$$\begin{aligned}\sum_m 2^{\theta m}\left|\sum_k x_{km}\right| &\le \sum_m 2^{\theta m} \sum_k |x_{km}| = \\ &= \sum_m \sum_k 2^{\theta k} 2^{\theta(m-k)} |x_{km}| \\ &= \sum_k 2^{\theta k}\left(\sum_m 2^{\theta(m-k)}|x_{km}|\right).\end{aligned}$$

We note that $J(2^{-k},u_k) = \max(\sum_m |x_{km}|, \sum 2^{m-k}|x_{km}|)$ since $2^{-k}|u_k|_{\lambda_1^1} = 2^{-k}\sum_m 2^m |x_{km}|$.

Now let us consider the function

$$f_k(t) = \sum 2^{t(m-k)}|x_{km}|.$$

It is easy to see that f_k is convex on $0 \le t \le 1$, so $f_k(\theta) \le \max(f_k(0), f_k(1)) = J(2^{-k},u_k)$. Hence, $x = \sum u_k \in \lambda_\theta^1$ and $\|x\|_{1,\theta} \le \|x\|_{J(\theta,1,\bar{\lambda}^1)}$. □

For $q>1$, we have natural maps

$$\lambda_\theta^q \xrightarrow{\ \alpha\ } J(\theta,q,\bar{\lambda}^1) \xrightarrow{\ \beta\ } J(\theta,q,\bar{\lambda}^\infty) \xrightarrow{\ \varphi\ } K(\theta,q,\bar{\lambda}^\infty) \xrightarrow{\ \delta\ } \lambda_\theta^q,$$

where $\alpha = \alpha_{\theta,q}$ is the map defined above, β is the map arising from the embedding of $\bar{\lambda}^1$ in $\bar{\lambda}^\infty$, φ is the standard natural transformation from $J(\theta,q,-)$ to $K(\theta,q,-)$, and $\delta = (\alpha_{\theta,q'})'$. Since the composition $\delta\circ\varphi\circ\beta\circ\alpha$ is the identity on λ_θ^q, this proves that λ_θ^q is an $L(\bar{\lambda}^1)$-module (even though we only have $\|Tx\|_{\theta,q} \le C_{\theta,q}\|T\|\|x\|_{\theta,q}$ for some $C_{\theta,q} \ge 1$). Thus, we are in a position to consider, as in Chapter VI, the left Kan extension of the functor $\{\overline{\lambda}^1\} \to \mathfrak{B}$ which "picks out" λ^q_θ. We recall that the left Kan extension was denoted by $\mathrm{Lan}_{\lambda^q_\theta}$ and is defined by

$$\mathrm{Lan}_{\lambda^q_\theta} \overline{X} = L(\overline{\lambda}^1, \overline{X}) \otimes_{L(\overline{\lambda}^1)} \lambda^q_\theta.$$

Let us abbreviate $\mathrm{Lan}_{\lambda^q_\theta}$ by $\mathrm{Lan}_{\theta,q}$. We can now prove the following result.

2.7. Theorem. $\mathrm{Lan}_{\theta,q}\overline{X} = J(\theta,q,\overline{X})$ for all $\overline{X} \in \overline{\mathfrak{B}}$.

Proof: We recall that $\mathrm{Lan}_{\theta,q}\overline{X}$ can be defined as the completion of $\Delta\overline{X}$ with respect to the (semi-)norm

$$\|x\|_{\mathrm{Lan}_{\theta,q}\overline{X}} = \inf\{\sum_1^n \|T_j\|\|a_j\|_{\theta,q} \mid T_j \in L(\overline{\lambda}^1, \overline{X}),\ a_j \in \Delta\overline{\lambda}^1,\ \Sigma\, T_j a_j = x\}.$$

We shall identify

$$\sum_1^n T_j \otimes a_j = \sum_1^n T_j \otimes \hat{a}_j(u) = ((\sum_1^n T_j a_j) \otimes u') \otimes u,$$

where u and u' denote the units of $\overline{\lambda}^1$ and $\overline{\lambda}^\infty$, respectively, given by $u_k = \begin{cases} 1 & k=0 \\ 0 & k\neq 0 \end{cases}$ and $u=u'$. Likewise, $J(\theta,q,\overline{X})$ is defined as the completion of $\Delta\overline{X}$ with respect to the (semi-)norm

$$\|x\|_{J(\theta,q,\overline{X})} = \inf\left\{\left(\sum_{-\infty}^{\infty} (2^{\theta k} J(2^{-k}, u_k))^q\right)^{1/q} \middle| u_k \in \Delta\overline{X},\ \sum_{-\infty}^{\infty} u_k = x\right\}.$$

Now let $x \in \Delta\overline{X}$ and suppose that $x = \sum_{-\infty}^{\infty} u_k$, where

$$\left(\sum_{-\infty}^{\infty} (2^{\theta k}J(2^{-k},u_k))^q\right)^{1/q} < (1+\varepsilon)\|x\|_{J(\theta,q,\overline{X})}.$$

Then we define $T \in L(\overline{\lambda}^1,\overline{X})$ by

$$T(e_k) = \begin{cases} \dfrac{u_k}{J(2^{-k},u_k)} & \text{if} \quad J(2^{-k},u_k) \neq 0 \\ 0 & \text{if} \quad J(2^{-k},u_k) = 0. \end{cases}$$

Then for $a = \{a_k\} \in \lambda_\theta^q$, where $a_k = J(2^{-k},u_k)$, we have $Ta = x$, $\|T\| = 1$, and

$$\|x\|_{\mathrm{Lan}_{\theta,q}\overline{X}} \leq \|a\|_{\theta,q} < (1+\varepsilon)\|x\|_{J(\theta,q,\overline{X})}.$$

Since this holds for every $\varepsilon>0$, it follows that

$$\|x\|_{\mathrm{Lan}_{\theta,q}\overline{X}} \leq \|x\|_{J(\theta,q,\overline{X})}.$$

Conversely, suppose we have $x = \sum_{j=1}^{n} T_j a_j$, where $\Sigma\|T_j\|\|a_j\|_{\theta,q} < (1+\varepsilon)\|x\|_{\mathrm{Lan}_{\theta,q}\overline{X}}$, for $T_j \in L(\overline{\lambda}^1,\overline{X})$ and $a_j \in \Delta\overline{\lambda}^1$. We may assume $\|T_j\| = 1$, so $T_j = \{x_{jk}\}_{k\in\mathbf{Z}}$, where $T_j(e_k) = x_{jk}$, and $\sup_k J(2^{-k},x_{jk}) \leq 1$. If $a_j = \{a_{jk}\}_{k\in\mathbf{Z}}$, we define $u_k = \sum_{j=1}^{n} a_{jk}x_{jk}$. Hence, $x = \sum_{-\infty}^{\infty} u_k$, and

$$\begin{aligned} \|x\|_{J(\theta,q,\overline{X})} &\leq \left[\sum_{-\infty}^{\infty} (2^{\theta k}J(2^{-k},u_k))^q\right]^{1/q} \\ &= \left[\sum_{-\infty}^{\infty} (2^{\theta k}J(2^{-k},\sum_{j=1}^{n} a_{jk}x_{jk}))^q\right]^{1/q} \\ &\leq \left[\sum_{-\infty}^{\infty} (2^{\theta k}\sum_{j=1}^{n} a_{jk}J(2^{-k},x_{jk}))^q\right]^{1/q} \end{aligned}$$

$$\leq \left[\sum_{-\infty}^{\infty} (2^{\theta k} \sum_{j=1}^{n} a_{jk})^q\right]^{1/q}$$

$$\leq \sum_{j=1}^{n} \left[\sum_{-\infty}^{\infty} (2^{\theta k} a_{jk})^q\right]^{1/q} = \sum_{j=1}^{n} \|a_j\|_{\theta,q}$$

$$< (1+\varepsilon)\|x\|_{Lan_{\theta,q}\overline{X}}.$$

This completes our proof. □

Thus, the $J(\theta,q)$-method of interpolation is a left Kan extension and, hence, is characterized by being the minimal functor F from $\overline{\mathcal{B}}$ to $\mathcal{B}$ such that $F(\overline{\lambda}^1) = \lambda_\theta^q$. Moreover, from the following fact that the J- and K-methods are dual functors of each other, we may deduce, using VII.5.6, that the $K(\theta,q')$-method is the maximal Aronszajn-Gagliardo functor sending $\overline{\lambda}^\infty$ to $\lambda_\theta^{q'}$.

2.8. Theorem. $DJ(\theta,q,-) = K(\theta,q',-)$ and $DK(\theta,q,-) = J(\theta,q',-)$ for $0 < \theta < 1$, $1 \leq q < \infty$.

Proof: We first recall that $DLan_{\theta,q}\overline{X} = L_{L(\overline{\lambda}^1)}(\lambda_\theta^q, \overline{X}\otimes\overline{\lambda}^1)$. Now if $x \in \Sigma\overline{X}$, then we observe that x defines an element of $L_{L(\overline{\lambda}^1)}(\Delta\overline{\lambda}^1, \overline{X}\otimes\overline{\lambda}^1)$, i.e. the element $a = \{a_k\} \longmapsto x\otimes a = \{a_k x\}$, in view of the description of $\overline{X}\otimes\overline{\lambda}^1$ in 2.5. Moreover, we have

$$\|x\otimes a\|_{\overline{X}\otimes\overline{\lambda}^1} = \sum_{-\infty}^{\infty} K(2^k, a_k x) = \sum_{-\infty}^{\infty} |a_k| K(2^k, x).$$

Therefore, recalling that $\{a_k\} \in \lambda_\theta^q$ when $(\sum_{-\infty}^{\infty} (2^{\theta k}|a_k|)^q)^{1/q} < \infty$, we see that x will define an element of $L_{L(\overline{\lambda}^1)}(\lambda_\theta^q, \overline{X}\otimes\overline{\lambda}^1)$ if and only if $(\Sigma(2^{-\theta k}K(2^k,x))^{q'})^{1/q'} < \infty$, i.e. if and only if $x \in K(\theta,q',\overline{X})$.

The fact that $DK(\theta,q,-) = J(\theta,q',-)$ follows from the equivalence of the J- and K-methods. In particular the natural

$\mathscr{B}_\infty$-isomorphism $\varphi: J(\theta,q,-) \to K(\theta,q,-)$ defined in II.3.1 also gives rise to a natural $\mathscr{B}_\infty$-isomorphism $D\varphi: DK(\theta,q,-) \to DJ(\theta,q,-)$. Since $DJ(\theta,q,-) = K(\theta,q',-)$, we see that the proper norming gives us $DK(\theta,q,-) = J(\theta,q',-)$. □

Finally, we can check that the functors $J(\theta,q,-)$ and $K(\theta,q,-)$ preserve the operations discussed in Chapter I of making a doolittle diagram into a Banach couple or "regularizing" a doolittle diagram. This tells us that the J- and K-methods are exactly the extensions of the classical methods that we would wish.

2.9. Theorem. For all $\overline{X} \in \overline{\mathscr{B}}$ we have

$$J(\theta,q,\overline{X}) = J(\theta,q,\overline{X}^0) = J(\theta,q,\overline{X}/\overline{K}) = J(\theta,q,\overline{X}^0/\overline{K})$$

and

$$K(\theta,q,\overline{X}) = K(\theta,q,\overline{X}^0) = K(\theta,q,\overline{X}/\overline{K}) = K(\theta,q,\overline{X}^0/\overline{K}).$$

Proof: The statements $J(\theta,q,\overline{X}) = J(\theta,q,\overline{X}^0)$, $J(\theta,q,\overline{X}/\overline{K}) = J(\theta,q,\overline{X}^0/\overline{K})$, $K(\theta,q,\overline{X}) = K(\theta,q,\overline{X}/\overline{K})$, and $K(\theta,q,\overline{X}^0) = K(\theta,q,\overline{X}^0/\overline{K})$ follow from Proposition VIII.1.1 in view of Theorem 2.7 (recall that $\overline{X}^0/\overline{K} = (\overline{X}/\overline{K})^0$ by I.2.2). Thus, the only statements that remain to be verified are $J(\overline{X}) = J(\overline{X}/\overline{K})$ and $K(\overline{X}) = K(\overline{X}^0)$. These were proved in Chapter VIII for $J = \mathrm{Lan}_A$ and K its dual (see VIII.2.5 and VIII.2.6) under certain conditions on $\overline{A}$. These conditions can be seen to hold for the regular Banach couple $\overline{A}=\overline{\lambda}^1$. In particular $\overline{\lambda}^1$ is quasi-projective with $\Delta\overline{\lambda}^1$ dense in λ_θ^q, while λ_θ^q is not properly contained in $\lambda_0^1 \cup \lambda_1^1$. Moreover, we can define $U,V \in L(\overline{\lambda}^1)$ satisfying the hypothesis of VIII.2.6 as translations: $U\{x_k\} = \{y_k\}$, where $y_k = x_{k-1}$, and $V=U^{-1}$. □

3. The Complex Method Revisited.

In addition to proving that the real J- and K-methods are minimal and maximal extensions, respectively, Janson [11] also proved a similar result for the complex methods. He formulated his theorem in terms of the discrete complex method (as defined by Cwikel) and worked with spaces of weighted Fourier transforms. An advantage of using Fourier transforms is that it is then possible to define a complex method also for real Banach spaces. In our setting, however, it is less natural to use Fourier transforms because the Fourier transform conceals the importance of using L^1-spaces for the generating diagrams.

We now introduce some diagrams and some interpolation spaces which will allow us to formulate our version of the complex methods and show that they can be viewed as left Kan extensions.

3.1. Definition. Let $\overline{A}$ denote the diagram

$$\overline{A} = \begin{array}{ccc} \Delta\overline{A} & \xrightarrow{\delta_0} & C_0(\mathbb{R}) \\ {\scriptstyle \delta_1}\downarrow & & \downarrow{\scriptstyle \sigma_0} \\ C_0(\mathbb{R}) & \xrightarrow[\sigma_1]{} & \Sigma\overline{A} \end{array} ,$$

where $C_0(\mathbb{R})$ is the space of continuous $\mathbb{C}$-valued functors on $\mathbb{R}$ which tend to 0 as $|t| \to \infty$ and $\Delta\overline{A} = A_0(S)$ is the space of bounded continuous functions f on the strip $S = \{z \in \mathbb{C} \mid 0 \le \mathrm{Re}\, z \le 1\}$ which are analytic in the interior, $\mathrm{int}(S)$, and such that

$f(k+it) \in C_0(\mathbb{R})$, $k=0,1$. The maps δ_k, $k=0,1$, are defined by

$$(\delta_k f)(t) = f(k+it).$$

The space $\Sigma\overline{A}$ does not have a simple description as a space of functions, but it may be described as the predual of the pullback, $H^1(S)$, in the dual diagram:

$$\overline{A}' = \begin{array}{ccc} H^1(S) & \xrightarrow{\sigma_0'} & M(\mathbb{R}) \\ \sigma_1' \big\downarrow & & \big\downarrow \delta_0' \\ M(\mathbb{R}) & \xrightarrow[\delta_1']{} & A_0(S)' \end{array},$$

where $H^1(S)$ is seen to be the space of functions f which are analytic on $\operatorname{int}(S)$ and are such that

$$\|f\|_{H^1} = \sup_{0<\theta<1} \int_{-\infty}^{\infty} |f(\theta+it)|\,dt < \infty.$$

The maps σ_k' and δ_k' behave as follows:

$$\langle \sigma_k' f, \varphi \rangle = \int_{-\infty}^{\infty} \varphi(t) f(k+it)\,dt,$$

for $f \in H^1(S)$ and $\varphi \in C_0(\mathbb{R})$,

$$\langle \delta_k'(\mu), f \rangle = \int_{-\infty}^{\infty} f(k+it)\,d\mu(t),$$

for $f \in A_0(S)$ and $\mu \in M(\mathbb{R})$. In particular, we see that for $f \in A_0(S)$, $h \in H^1(S)$,

$$\begin{aligned}\langle \delta_0' \sigma_0'(h), f\rangle &= \int_{-\infty}^{\infty} h(it)f(it)\,dt \\ &= \int_{-\infty}^{\infty} h(1+it)f(1+it)\,dt \\ &= \langle \delta_1' \sigma_1'(h), f\rangle = \langle j'(h), f\rangle.\end{aligned}$$

We note that $\bar{A}'$ is not regular. In many cases $\bar{A}'$ can be replaced by the diagram $(\bar{A}')^0$, which we shall call $\bar{H}$. Explicitly, $\bar{H}$ is given by

$$\bar{H} = \begin{array}{ccc} H^1(S) & \xrightarrow{\sigma_0'} & L^1(\mathbb{R}) \\ \sigma_1' \big\downarrow & & \big\downarrow \delta_0' \\ L^1(\mathbb{R}) & \xrightarrow[\delta_1']{} & \Sigma\bar{H}. \end{array}$$

Again $\Sigma\bar{H}$ is best described as the predual of the pullback, $H^\infty(S)$, in the following dual diagram

$$\bar{H}' = \begin{array}{ccc} H^\infty(S) & \xrightarrow{\delta_0''} & L^\infty(\mathbb{R}) \\ \delta_1'' \big\downarrow & & \big\downarrow \sigma_0'' \\ L^\infty(\mathbb{R}) & \xrightarrow[\sigma_1'']{} & H^1(S)', \end{array}$$

where $H^\infty(S)$ is the space of functions f analytic on $\mathrm{int}(S)$ such that $\sup\{|f(z)|\, z \in \mathrm{int}(S)\} < \infty$. We note that δ_k'' and σ_k'' are essentially the δ_k and σ_k in the diagram $\bar{A}$ except that they are defined on bigger spaces.

We shall soon see how the diagrams $\bar{A}$, $\bar{A}'$, $\bar{H}$, and $\bar{H}'$ are connected to the complex methods. First we observe that $\bar{H}$ and $\bar{A}'$ are compactly quasi-projective, while $\bar{A}$ is compactly quasi-injective and $\bar{H}'$ is quasi-injective (see Definitions VIII.2.1). Thus, we expect the spaces $L(\overline{H},\overline{X})$, $L(\overline{A}',\overline{X})$, $L(\overline{X},\overline{A})$, and $L(\overline{X},\overline{H}')$ to have simple representations.

In order to describe the above spaces we need some new notation and some general facts.

First we shall write $H_\infty(S,X)$, for any $X\in\mathcal{B}$, to denote the space of X-valued functions analytic on $\operatorname{int}(S)$ such that $\sup\{\|f(z)\|_X \mid z \in \operatorname{int}(s)\} < \infty$. It is known that functions in $H_\infty(S,X)$ need not have boundary values in X. However, for every $z \in \operatorname{int}(S)$, f can be interpreted as an element T_z of $L(L^1(\mathbb{R}),X)$ defined by

$$T_z(\varphi) = \int_{-\infty}^{\infty} f(z+it)\varphi(t)dt.$$

Then, as proved by Peetre [20], f has boundary values (at least in the weak operator topology) in the space $L(L^1(\mathbb{R}),X)$. Under the condition that X satisfy the Radon-Nikodym property, we have

$$L(L^1(\mathbb{R}),X) = L^\infty(\mathbb{R},X);$$

in fact the inclusion $L^\infty(\mathbb{R},X) \subset L(L^1(\mathbb{R}),X)$ is reversible if and only if X has the Radon-Nikodym property [7]. In the following definition we shall use the notation of Peetre [20].

3.2. Definitions. Let $\overline{X}$ be a Banach couple. We define

$$H^\infty(S,\overline{X}) = \{f \in H_\infty(S,\Sigma\overline{X}) \mid f(k+it) \in L^\infty(\mathbb{R},X_k),\ k=0,1\}$$

and

$$H_\infty(S,\overline{X}) = \{f \in H_\infty(S,\Sigma\overline{X}) \mid f(k+it) \in L(L^1(\mathbb{R}),X_k),\ k=0,1\}.$$

3.3. Remark. From the above comments we see that if X_0 and X_1

have the Radon-Nikodym property, then $H^{\infty}(S,\overline{X}) = H_{\infty}(S,\overline{X})$.

Now we proceed to describe $L(\overline{H},\overline{X})$.

3.4. <u>Proposition</u>. Let $\overline{H} = (\overline{A}')^{0}$ be as above, let $\overline{X}\in\mathfrak{B}$, and let $q:\overline{X} \to \overline{X}/\overline{KX}$ be the canonical projection map. Then an element T of $L(\overline{H},\overline{X})$ can be represented as a triple (T_0,T_1,f_T), where $T_k \in L(L^1(\mathbb{R}),X_k)$, $k=0,1$, and where $f_T \in H_{\infty}(S,\overline{X}/\overline{KX})$ is such that, as an element of $L(L^1(\mathbb{R}),\Sigma(\overline{X}/\overline{KX}))$, we have

$$\int_{-\infty}^{\infty} f_T(k+it)\varphi(t)dt = q_k\circ T_k(\varphi), \quad \varphi \in L^1(\mathbb{R}).$$

<u>Proof</u>: Since $T' \in L(\overline{X}',\overline{H}')$, we have the map $T':\Delta\overline{X}' \to \Delta\overline{H}' = H^{\infty}(S)$. For $z \in \operatorname{int}(S)$ we define f_T by the relation

$$\langle f_T z,x'\rangle = \langle T'(x'),z\rangle$$

for $x' \in \Delta\overline{X}' = (\Sigma\overline{X})'$. It follows that $f_T(z) \in (\Sigma\overline{X})''$. We want to prove actually that $f_T(z) \in \Sigma\overline{X}$. To do this we shall use a Fourier transform argument. We consider the functions $h_{\xi}(z) = \exp(z^2)\exp(-\xi z)$, which are in $H^{\infty}(S)$. Then we have for every $x'\in\Delta\overline{X}'$,

$$\begin{aligned}\langle x',j(Th_{\xi})\rangle &= \langle x',\sigma_0\circ T_0\circ\delta_0(h_{\xi})\rangle\\ &= \int_{-\infty}^{\infty}\langle x',f_T(it)\rangle e^{-t^2}e^{-i\xi t}dt\\ &= \langle x',\sigma_1\circ T_1\circ\delta_1(h_{\xi})\rangle\\ &= \int_{-\infty}^{\infty}\langle x',f_T(1+it)\rangle e^{(1+it)^2}e^{-\xi(1+it)}dt.\end{aligned}$$

Now $|\langle x',f_T\rangle| \le \|x'\|\|T\|$, so

$$|\langle x', j(Th_\xi)\rangle| = |\langle x', \sigma_0 \circ T_0 \circ \delta_0(h_\xi)\rangle|$$
$$\le \|x'\|\|T\| \int_{-\infty}^{\infty} |e^{-t^2} e^{-i\xi t}| dt = \|x'\|\|T\|\sqrt{\pi}.$$

However, we also have that

$$|e^{\xi}\langle x', j(Th_\xi)\rangle| = |e^{\xi}\langle x', \sigma_1 \circ T_1 \circ \delta_1(h_\xi)\rangle|$$
$$= |\int_{-\infty}^{\infty} \langle x', f_T(1+it)\rangle e^{(1+it)^2} e^{-i\xi t} dt|$$
$$\le \|x'\|\|T\| \int_{-\infty}^{\infty} e^{(1+it)^2} e^{-i\xi t} dt = \|x'\|\|T\| e\sqrt{\pi}.$$

Therefore, for every $x' \in \Delta\overline{X}'$, $\|x'\| \le 1$, we have

$$\langle x', j(Th_\xi)\rangle \le \sqrt{\pi}\ \|T\|$$

and

$$|e^{\xi}\langle x', j(Th_\xi)\rangle| \le e\sqrt{\pi}\ \|T\|.$$

Now we define $\hat{g}(\xi) = j(Th_\xi)$. We see that both $\hat{g}(\xi)$ and $e^{\xi}\hat{g}(\xi)$ are in $C_b(\mathbb{R}, \Sigma\overline{X})$, where C_b denotes the bounded continuous functions. Hence, for every s, $0 < s < 1$, $e^{s\xi}\hat{g}(\xi) \in L^1(\mathbb{R}, \Sigma\overline{X})$. Thus, for every $z = s + it$, $0 < s < 1$, we have

$$e^{z^2} f_T(z) = 1/2\pi \int_{-\infty}^{\infty} \hat{g}(\xi) e^{s\xi} e^{i\xi t} d\xi \in \Sigma\overline{X}.$$

Since we also have that for every $x' \in \Delta\overline{X}'$

$$|\langle x', f_T(it)\rangle| \le \|\sigma_0'(x')\|_{X_0'} \|T\|$$

and

$$|\langle x', f_T(1+it)\rangle| \le \|\sigma_1^!(x')\|_{X_1^!}\|T\|,$$

it follows that the boundary values of f_T lie in $L(L^1(\mathbb{R}), X_1/K_1\bar{X})$, $k=0,1$, and, furthermore, that

$$\int_{-\infty}^{\infty} \langle x', f_T(k+it)\rangle \varphi(t)dt = \langle x', T_k(\varphi)\rangle,$$

for $\varphi \in L^1(\mathbb{R})$. □

Given the description of the space $L(\bar{H},\bar{X})$, we can find simple descriptions as well of the spaces $L(\bar{X},\bar{H}')$, $L(\bar{X},\bar{A})$, and $L(\bar{A}',\bar{X})$. First we see that quite simply

$$L(\bar{X},\bar{H}') = L(\bar{H},\bar{X}') = H_\infty(S,(\bar{X}/\bar{K}\bar{X})').$$

Second, we have $L(\bar{X},\bar{A}) \subset L(\bar{X},\bar{H}') = L(\bar{H},\bar{X}')$, and, more precisely, $L(\bar{X},\bar{A})$ consists of triples (T_0,T_1,f_T), where $T_k \in L(X_k,C_0(\mathbb{R}))$, and $f_T \in H_\infty(S,\bar{X}'/\bar{K}\bar{X}')$ are such that $f_T(k+it) = q_k \circ T_k(t)$, $k=0,1$. Third, the space $L(\bar{A}',\bar{X})$ is relatively uninteresting since

$$\bar{A}' = \bar{H}\sqcup\bar{M}_S,$$

where

$$\bar{M}_S = \begin{array}{ccc} \{0\} & \longrightarrow & M_S(\mathbb{R}) \\ \downarrow & & \downarrow \\ M_S(\mathbb{R}) & \longrightarrow & M_S(\mathbb{R})\sqcup M_S(\mathbb{R}), \end{array}$$

and $M_S(\mathbb{R}) = \{\mu \in M(\mathbb{R}) \mid \mu \perp dt\}$. Therefore,
$L(\overline{A}',\overline{X}) = L(\overline{H},\overline{X}) \sqcap L(M_S(\mathbb{R}),X_0) \sqcap L(M_S(\mathbb{R}),X_1)$.

Now we shall apply the complex methods of interpolation to the diagrams $\overline{A}$ and $\overline{H}$. Once we determine what the spaces $C_\theta\overline{A}$, $C_\theta\overline{H}$, and $C^\theta\overline{H}$ are, we shall be able to show that C_θ and C^θ are left Kan extensions, C_θ on all of $\overline{\mathcal{B}}$ and C^θ only on $\overline{\mathcal{BC}}$.

We recall that $C_\theta\overline{X}$ and $C^\theta\overline{X}$ are defined by $C_\theta\overline{X} = A(S,\overline{X})/I(\theta)$ and $C^\theta\overline{X} = \Lambda(S,\overline{X})/K(\theta)$, where $A(S,\overline{X})$ and $\Lambda(S,\overline{X})$ are certain spaces of $\Sigma\overline{X}$-valued functions analytic on $\mathrm{int}(S)$ (see II.2).

3.5. <u>Proposition</u>. $C_\theta\overline{A} = C_0(\mathbb{R})$, $C_\theta\overline{H} = L^1(\mathbb{R})$, and $C^\theta\overline{H} = M(\mathbb{R})$. Furthermore, the maps $\delta_\theta:\Delta\overline{X} \to C_\theta\overline{X}$ and $\sigma_\theta:C_\theta\overline{X} \to \Sigma\overline{X}$, for $\overline{X}=\overline{A}$ or $\overline{H}$, are given by $\delta_\theta(f)(t) = f(\theta+it)$, $f\in\Delta\overline{X}$, and $\langle x',\sigma_\theta(f)\rangle = \langle\delta_\theta(x'),f\rangle$, $f\in C_\theta\overline{X}$, $x'\in\Delta\overline{X}'$.

<u>Proof</u>: Since the proposition is all but obvious we shall only indicate the proof. The main point is that there exist maps

$$\varphi:C_0(\mathbb{R}) \to A(S,\overline{A}) \subset L(\overline{H},\overline{A}),$$
$$\psi:L^1(\mathbb{R}) \to A(S,\overline{H}) \subset L(\overline{H},\overline{H})$$

given by $(\varphi g)(h) = h*g$ and $(\psi g)(h) = h*g$, for g in $C_0(\mathbb{R})$ or $L^1(\mathbb{R})$, respectively, and $h \in \Delta\overline{H} = H^1(S)$. This shows that $C_0(\mathbb{R}) \subset C_\theta\overline{A}$ and $L^1(\mathbb{R}) \subset C_\theta\overline{H}$. By a duality argument, one has $C_\theta\overline{A} \subset C_0(\mathbb{R})$ and $C_\theta\overline{H} \subset L^1(\mathbb{R})$. Finally,

$$C^\theta(\overline{H}) = C^\theta((\overline{A}')^0) = C^\theta(\overline{A}') = (C_\theta\overline{A})' = M(\mathbb{R}). \qquad \square$$

3.6. <u>Theorem</u>. 1. $C_\theta\overline{X} = \mathrm{Lan}_{C_\theta\overline{H}}\overline{X}$ for all $\overline{X}\in\overline{\mathcal{B}}$. 2. $C^\theta\overline{X} = \mathrm{Lan}_{C^\theta\overline{H}}\overline{X}$ for all $\overline{X}\in\overline{\mathcal{BC}}$.

Proof: First we shall prove (2). When $\overline{X}$ is a Banach couple, we know that $\Lambda(S,\overline{X})$ and $H_\infty(S,\overline{X})$ are isometrically isomorphic (see [20]) and, moreover, that the maps

$$h_\theta : H_\infty(S,\overline{X}) \to \Sigma\overline{X},$$
$$d_\theta : \Lambda(S,\overline{X}) \to \Sigma\overline{X}$$

given by $h_\theta(f) = f(\theta)$, $d_\theta(f) = f'(\theta)$ are isomorphic as maps. Hence, it suffices to prove that $L(\overline{H},\overline{X})\otimes_{L\overline{H}}C^\theta\overline{H}$ $(=\mathrm{Lan}_{C^\theta\overline{H}}\overline{X})$ and $H_\infty(S,\overline{X})/I(\theta)$ $(I(\theta) = \ker(h_\theta))$ are isomorphic. Consider $T\otimes\mu \in L(\overline{H},\overline{X})\otimes_{L\overline{H}}C^\theta\overline{H}$, recalling from Proposition 3.5 that $C^\theta\overline{H} = M(\mathbb{R})$. Then since $\mu = \mu*\varepsilon$ (where ε is the ordinary Dirac measure concentrated at 0), we have

$$T\otimes\mu = T\otimes\mu*\varepsilon = T*\mu\otimes\varepsilon .$$

By means of the map $\sigma_\theta : C^\theta\overline{H} \to \Sigma\overline{H}$, we can define an evaluation map

$$ev_\theta : L(\overline{H},\overline{X})\otimes_{L\overline{H}}C^\theta\overline{H} \to \Sigma\overline{X}$$

by

$$ev_\theta(T\otimes\varepsilon) = T(\sigma_\theta(\varepsilon)) = f_T(\theta);$$

so

$$ev_\theta(T\otimes\mu) = ev_\theta(T*\mu\otimes\varepsilon) = f_{T*\mu}(\theta).$$

Now we have to show that if $ev_\theta(T\otimes\varepsilon) = 0$, i.e. if $f_T(\theta) = 0$, then $T\otimes\varepsilon = 0$ in $L(\overline{H},\overline{X})\otimes_{L\overline{H}}C^\theta\overline{H}$. Then in view of our characterization of $C^\theta\overline{X}$ and the fact that $f_T \in H_\infty(S,\overline{X})$, the result will follow. To this end, consider the function

$$\alpha(z) = \frac{e^{\pi iz} - e^{\pi i\theta}}{e^{\pi iz} - e^{-\pi i\theta}}$$

and write $f_T(z) = \alpha(z)\cdot 1/\alpha(z)\cdot f_T(z)$. If we can prove that $1/\alpha\cdot f_T \in H_\infty(S,\overline{X})$, then $1/\alpha\cdot f_T = f_U$ for some $U \in L(\overline{H},\overline{X})$ and, hence, $T\otimes\varepsilon = U\alpha\otimes\varepsilon = U\otimes(\alpha\cdot\varepsilon) = U\otimes 0 = 0$. To show that $1/\alpha\cdot f_T \in H_\infty(S,\overline{X})$, consider $x'\in\Delta\overline{X}'$. Then $\langle x',1/\alpha\cdot f_T\rangle = 1/\alpha\langle x',f_T\rangle \in H_\infty(S,\mathbb{C})$, and $\langle x',1/\alpha\cdot f_T\rangle \le \|x'\|\|f_T\|$ since $|\alpha(z)| = 1$ if $\mathrm{Re}(z) = 0$ or 1. Hence, $1/\alpha\cdot f_T \in H_\infty(S,\Sigma\overline{X})$, which completes the proof of (2).

To prove (1) we first consider the case when $\overline{X}$ is a Banach couple. Let $T\otimes g \in L(\overline{H},\overline{X})\otimes_{L\overline{H}}C_\theta\overline{H} = \mathrm{Lan}_{C_\theta\overline{H}}\overline{X}$. Then we have

$$T\otimes g = T\otimes g*\varepsilon = T*g\otimes\varepsilon$$

and we have a quotient map

$$q: L(\overline{H},\overline{X})\otimes_{L^1\mathbb{R}}C_\theta\overline{H} \to L(\overline{H},\overline{X})\otimes_{L\overline{H}}C_\theta\overline{H}.$$

Now if $f_T \in H_\infty(S,\overline{X})$ and $g \in L^1(\mathbb{R}) = C_\theta\overline{H}$, then we see that $f_T*g \in A(S,\overline{X})$ since it has continuous, even uniformly continuous, boundary values. In particular, it follows that $L(\overline{H},\overline{X})\otimes_{L^1(\mathbb{R})} L^1(\mathbb{R}) \subset A(S,\overline{X})$. However, $L^1(\mathbb{R})$ is a Banach algebra with an approximating unit, and $A_0(S,\overline{X})$ is continuous under translations. Therefore, by the Cohen-Hewitt factorization theorem (see [5])

$$A_0(S,\overline{X})\otimes_{L^1(\mathbb{R})} L^1(\mathbb{R}) = A_0(S,\overline{X}).$$

Thus, we have

$$A_0(S,\bar{X}) \subset L(\bar{H},\bar{X}) \otimes_{L^1\mathbb{R}} L^1(\mathbb{R}) \subset A(S,\bar{X})$$

and since $A_0(S,\bar{X})/I(\theta) = A(S,\bar{X})/I(\theta) = C_\theta(\bar{X})$, we obtain

$$(L(\bar{H},\bar{X}) \otimes_{L^1(\mathbb{R})} L^1(\mathbb{R}))/I(\theta) = C_\theta\bar{X},$$

which proves (1) when $\bar{X}$ is a Banach couple.

If $\bar{X}$ is not a Banach couple, then we can use the fact that $\bar{H}$ is compactly quasi-projective so that for every $f \in A_0(S,\bar{X}/\bar{K}\bar{X})$, we can find a lifting of the continuous boundary functions f_0 and f_1 into X_0 and X_1, respectively, and this implies that the natural quotient $q_*: A_0(S,\bar{X}) \to A_0(S,\bar{X}/\bar{K}\bar{X})$ is onto. Therefore

$$C_\theta\bar{X} = A_0(S,\bar{X}/\bar{K}) = L(\bar{H},\bar{X}) \otimes_{L\bar{H}} C_\theta\bar{H}. \qquad \square$$

The above theorem makes it natural to ask if $C^\theta(\bar{X}/\bar{K}\bar{X}) = C^\theta\bar{X}$ is also $L(\bar{H},\bar{X}) \otimes_{L\bar{H}} C^\theta\bar{H}$ for <u>any</u> doolittle diagram $\bar{X}$. We may answer this question in the negative. To do this, let us write $L^1(\mathbb{R})$ as a quotient of a weighted ℓ^1-space $\ell^1(\Gamma,\omega)$, where $\Gamma = L^1(\mathbb{R})$ as a set and for the sequence $x = \{x_f\}_{f\in\Gamma}$ we define $\|x\| = \Sigma|x_f|\|f\|$. We may obtain a doolittle diagram $\ell\bar{H}$ by forming the pullback of the following diagram:

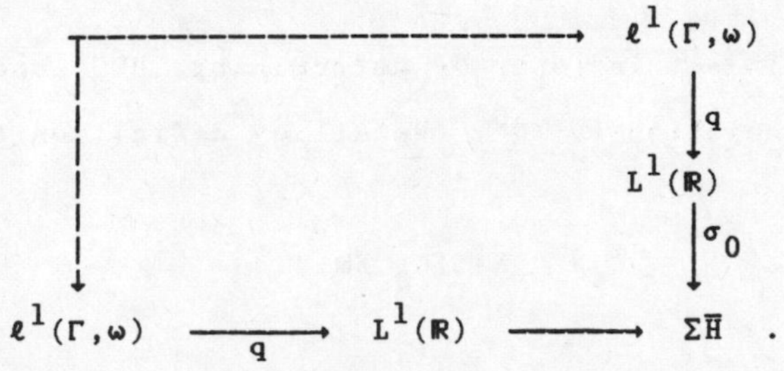

We then can see that $\ell\bar{H}/\bar{K}\ell\bar{H} = \bar{H}$, so that

$C^\theta(\ell\overline{H}) = C^\theta\overline{H} = L(\overline{H},\overline{H})\otimes_{L\overline{H}}C^\theta\overline{H}$. Thus we shall show that $C^\theta(\ell\overline{H}) \neq L(\overline{H},\ell\overline{H})\otimes_{L\overline{H}}C^\theta\overline{H}$ by showing that $\mathrm{Lan}_{C^\theta\overline{H}}(\ell\overline{H}) = C_\theta\overline{H} \neq C^\theta\overline{H}$. First, we see that if $f \in L^\infty(\mathbb{R},\ell^1(\Gamma,\omega)\Pi\ell^1(\Gamma,\omega))$, then $q\circ f \in L^\infty(\mathbb{R},L^1(\mathbb{R})\Pi L^1(\mathbb{R}))$. Now this means that if (T_0,T_1,f_T) is the representation of $f \in L(\overline{H},\ell\overline{H})$, then $f_T \in H^\infty(S,\overline{H})$. However, $H^\infty(S,\overline{H})/I(\theta) = C_\theta\overline{H}$, as proved by Peetre [20], which establishes our claim.

This same example also shows that $DC_\theta\overline{X} \neq \mathrm{Lan}_{C^\theta\overline{H}}\overline{X}$ since using the results of Chapter VIII, we have

$DC_\theta(\ell\overline{H}) = DC_\theta\overline{H} = DC_\theta((\overline{A}')^0) = DC_\theta\overline{A}' = (C_\theta\overline{A})' = C^\theta(\overline{A}') = C^\theta\overline{H}$.

3.7. <u>Remark</u>. We have previously avoided defining $\mathrm{Lan}_A\overline{X}$ when A is not a Δ-interpolation space. However, since $L(\overline{H},\overline{X})\otimes C^\theta\overline{H} = \mathrm{Lan}_{C^\theta\overline{H}}$ is easy to determine and turns out to be a Σ-interpolation space, we have violated our own rules. Nevertheless, the example above shows that $\mathrm{Lan}_{C^\theta\overline{H}}$ is not really a good functor, while the functor C_θ is natural and good also in our setting.

4. <u>The Dual Functor of C_θ</u>.

We shall conclude this paper by determining DC_θ and investigating its relation to C^θ. Recall by definition that

$$DC_\theta\overline{X} = \mathrm{NAT}(C_\theta,\overline{X}\otimes -).$$

We can use the natural maps $t \mapsto t_{\overline{I}}$ and $t \mapsto t_{\overline{X}'}$ from $DC_\theta\overline{X}$ to $L_{L\overline{I}}(C_\theta\overline{I},\overline{X}\otimes\overline{I}) = \Sigma\overline{X}$ and $L_{L\overline{X}'}(C_\theta\overline{X}',\overline{X}\otimes\overline{X}')$, respectively, to obtain the maps

$$\begin{array}{ccc} DC_\theta\overline{X} & \longrightarrow & L_{L\overline{X}'}(C_\theta\overline{X}',\overline{X}\otimes\overline{X}') \to L_{L\overline{X}'}(C_\theta\overline{X}',\overline{X}''\otimes\overline{X}') = C^\theta\overline{X}'' \\ \downarrow & & \\ \Sigma\overline{X} & & \end{array}$$

Since both of these maps are injective, we have $DC_\theta\overline{X} \subset \Sigma\overline{X} \cap C^\theta\overline{X}''$. Conversely, if $x\in\Sigma\overline{X} \cap C^\theta\overline{X}''$, then for every $\overline{Y}\in\mathfrak{B}$ and $y\in\Delta\overline{Y}$, $x\otimes y \in \overline{X}\otimes\overline{Y}$ and

$$\begin{aligned}\|x\otimes y\|_{\overline{X}\otimes\overline{Y}} &= \sup\{|\langle Ty,x\rangle| \;\big|\; T \in L(\overline{Y},\overline{X}'),\|T\| \le 1\} \\ &\le \|Ty\|_{C_\theta\overline{X}'}\|x\|_{(C_\theta\overline{X}')'} = \|Ty\|_{C_\theta\overline{X}'}\|x\|_{C^\theta\overline{X}''}.\end{aligned}$$

But $\|Ty\|_{C_\theta\overline{X}'} \le \|T\|\|y\|_{C_\theta\overline{Y}}$, so $x \in L_{L\overline{Y}}(C_\theta\overline{Y},\overline{X}\otimes\overline{Y})$. Thus, we have obtained the following characterization.

4.1. <u>Proposition</u>. $DC_\theta\overline{X} = \Sigma\overline{X} \cap C^\theta\overline{X}''$.

Furthermore, we have the following proposition.

4.2. <u>Proposition</u>. $DC_\theta\overline{X} = DC_\theta(\overline{X}/\overline{K\overline{X}})$.

<u>Proof</u>: Since $C_\theta\overline{X} = \mathrm{Lan}_{C_\theta\overline{H}}\overline{X}$, we can use VII.2.5 to obtain

$$DC_\theta\overline{X} = L_{L\overline{H}}(C_\theta\overline{H},\overline{H}\otimes\overline{X}).$$

Moreover, since $\overline{H}$ is regular, it follows from VIII.1.1 that $\overline{H}\otimes\overline{X} = \overline{H}\otimes(\overline{X}/\overline{K\overline{X}})$ and, therefore, $DC_\theta\overline{X} = DC_\theta(\overline{X}/\overline{K\overline{X}})$. □

It remains to decide if $DC_\theta\overline{X}$ is equal to $C^\theta\overline{X}$. However, the question that really needs to be answered is whether $DC_\theta\overline{X}$ equals $C^\theta\overline{X}$ for doolittle diagrams $\overline{X}$ which are not duals since there <u>is</u> agreement on dual diagrams as we see below.

4.3. <u>Theorem</u>. $DC_\theta\overline{X}' = C^\theta\overline{X}'$ for all $\overline{X}\in\overline{\mathfrak{B}}$.

<u>Proof</u>: We can see this result in two ways. First, one can combine the results 3.6, VII.3.3, and III.2.6 to obtain

$$DLan_{C_\theta\overline{H}}\overline{X}' = (Lan_{C_\theta\overline{H}}\overline{X})' = (C_\theta\overline{X})' = C^\theta\overline{X}',$$

since $\overline{H}$ is a regular couple satisfying the metric approximation property. Alternatively, one can use Proposition 4.1 to get a more direct proof. □

It is partly, however, the characterization of $DC_\theta\overline{X}$ in Proposition 4.1 that makes it seem unlikely that DC_θ and C^θ agree in general. Moreover, although we have not been able to decide whether this is true, we also do not feel that $H_\infty(S,\overline{X})$ is a natural H^∞-space since its elements are not determined by their behaviour on int(S). Thus, we wish to define a new functor which we believe is the correct functor for DC_θ.

To motivate the definition of our functor, we start by recalling from Proposition III.2.3 that C_θ is an exact interpolation functor of exponent θ and, hence, that for $x\in\Delta\overline{X}$,

$$\|x\|_{C_\theta\overline{X}} \le \|\delta_0x\|_{X_0}^{1-\theta}\|\delta_1x\|_{X_1}^{\theta}.$$

This means that the C_θ-norm is dominated by the geometric means of the norms $\|\ \|_{X_0}$, $\|\ \|_{X_1}$. Since the geometric mean is dominated by the arithmetic mean, it follows that if $f \in A(S,\overline{X})$, then

$$\|f(s+it)\|_{A(S,\overline{X})} \le (1-s)\|f_0\|_{X_0} + s\|f_1\|_{X_1}.$$

Now $A(s,x) = (1-s)\|x\|_{X_0} + s\|x\|_{X_1}$ is a norm on $\Delta\overline{X}$ with dual norm on $\Sigma\overline{X}'$ given by

$$M(s,x') = \inf\{\max((1/1-s)\|x_0'\|_{X_0'},\ (1/s)\|x_1'\|_{X_1'})\,|\,x_0' + x_1' = x'\}.$$

Likewise, we have for $x \in \Sigma\overline{X}$,

$$\begin{aligned} M(s,x) &= \inf\{\max(1/1-s)\|x_0\|_{X_0},\ (1/s)\|x_1\|_{X_1}\,|\,x_0 + x_1 = x\} \\ &= \sup\{|\langle x',x\rangle|\,|\,x' \in \Delta\overline{X}',\ (1-s)\|\sigma_0'x'\|_{X_0'} + s\|\sigma_1'x'\|_{X_1'} \le 1\}. \end{aligned}$$

Therefore, the following definition becomes somewhat natural.

4.4. <u>Definition</u>. Let $H_*^\infty(S,\overline{X})$ be the space of all $\Sigma\overline{X}$-valued functions f which are analytic on $\operatorname{int}(S)$ and are such that

$$\sup\{M(s,f(s+it))\,|\,s+it \in \operatorname{int}(S)\} < \infty.$$

4.5. <u>Conjecture</u>. $DC_\theta\overline{X} = H_*^\infty(S,\overline{X})/I(\theta)$, where $I(\theta) = \ker(h_\theta)$ and $h_\theta : H_*^\infty(S,\overline{X}) \to \Sigma\overline{X}$ is given by $h_\theta(f) = f(\theta)$.

In support of this conjecture, we may say that the space $H_*^\infty(S,\overline{X})$ appears to be the maximal H^∞-space possible for interpolation and the following proposition indicates that it does have good properties.

4.6. <u>Proposition</u>. (i) If $f \in H_\infty(S,\overline{X})$, then $\log M(s,f(s+it))$ is subharmonic on $\operatorname{int}(S)$ and $\sup M(s,f(s+it)) \le \|f\|_{H_\infty(S,\overline{X})}$, which implies that $H_\infty(S,\overline{X}) \subset H_*^\infty(S,\overline{X})$. (ii) Conversely, if $f \in H_\infty(S,\Sigma\overline{X})$

and if $\sup M(s,f(s+it)) \le 1$, then the boundary functions $f_k(t) = f(k+it)$, $k=0,1$, map the unit ball of $L^1(\mathbb{R})$ into the closure (in $\Sigma\overline{X}$) of the unit ball of $X_k/K_k\overline{X}$. Hence, $f \in H_\infty(S,(\overline{X}/\overline{K}\overline{X})^\sim)$, where $(\ )^\sim$ denotes the Gagliardo completion (see [2] or [11]).

<u>Proof</u>: (i) It follows from the definition of $M(s,x)$ that

$$M(s,f(s+it)) = \sup\left\{\frac{\langle x',f(s+it)\rangle}{A(s,x')} \;\middle|\; x'\neq 0,\ x'\in\Delta\overline{X}'\right\}.$$

Therefore,

$$\log M(s,f(s+it)) = \sup\{\log|\langle x',f(s+it)\rangle| - \log A(s,x')\},$$

so it suffices to prove that for every non-zero $x'\in\Delta\overline{X}'$, $\log|\langle x',f(s+it)\rangle| - \log A(s,x')$ is subharmonic. However, $\langle x',f(s+it)\rangle$ is an ordinary analytic function, so $\log|\langle x',f(s+it)\rangle|$ is clearly subharmonic. Furthermore, $A(s,x') = (1-s)\|\sigma_0'x'\|_{X_0} + s\|\sigma_1'x'\|_{X_1}$, so

$$-\log A(s,x') = -\log((1-s)\|\sigma_0'x'\|_{X_0} + s\|\sigma_1'x'\|_{X_1})$$

is a convex function of s and is therefore also subharmonic. Now, using the Poisson integral, we have

$$f(s+it) = \int f(iy)P_0(s+it,y)dy + \int f(1+iy)P_1(s+it,y)dy,$$

as a representation, $x_0 + x_1$. However, $P_k\in L^1(\mathbb{R})$, $k=0,1$, and $\int P_k(s+it,y)dy = |k+s-1|$, so $\|x_0\| \le (1-s)\|f_0\|$ and $\|x_1\| \le s\|f_1\|$, i.e. $M(s,f(s+it)) \le \|f\|_{H_\infty(S,\overline{X})}$.

To prove (ii) we observe that since $f \in H_\infty(S,\Sigma\overline{X})$, we have boundary values in $L(L^1(\mathbb{R}),\Sigma\overline{X})$. Therefore, let $g \in L^1(\mathbb{R})$, $\|g\| \le 1$, and consider $f*g$ defined by

$$F_g(z) = f*g(z) = \int_{-\infty}^{\infty} f(z+it)g(-t)dt.$$

F_g is continuous on S, so $F_g \in A(S,\overline{X})$. Furthermore, since $M(s,x)$ is a norm on $\Sigma\overline{X}$, we have

$$M(s,F_g(s+it)) \le \sup_{-\infty<y<\infty} M(s,f(s+iy)).$$

Therefore, if $x' \in \Delta\overline{X}'$, we have

$$|\langle x',F_g(s+it)\rangle| \le (1-s)\|\sigma_0' x'\|_{X_0'} + s\|\sigma_1' x'\|_{X_1'} ,$$

so in particular, $|\langle x',F_g(it)\rangle| \le \|\sigma_0' x'\|_{X_0'}$ and $|\langle x',F_g(1+it)\rangle| \le \|\sigma_1' x'\|_{X_1'}$. Thus, $F_g(k+it)$ belongs to the (weak) closure of the unit ball of X_k, $k=0,1$, and this proves (ii). □

4.7. <u>Corollary</u>. If $\overline{X}$ is Gagliardo closed, then $H_*^{\infty}(S,\overline{X}) = H_{\infty}(S,\overline{X})$. In particular, $H_*^{\infty}(S,\overline{X}') = H_{\infty}(S,\overline{X}')$ for all $\overline{X} \in \overline{\mathfrak{B}}$.

4.8. <u>Remark</u>. The above corollary shows that our conjecture leads to no conflict with known results on DC_θ since

$$DC_\theta(\overline{X}') = C^\theta\overline{X}' = H_\infty(S,\overline{X}')/I(\theta) = H_*^{\infty}(S,\overline{X}')/I(\theta).$$

With these new ideas and the above conjecture and the sincere wish that they will be pursued by those interested in interpolation theory, we end our exposition.

BIBLIOGRAPHY

1. Aronszajn, N. and Gagliardo, E., Interpolation spaces and interpolation methods, Ann. Mat. Pura Appl. 68 (1965), 51-118.
2. Bergh, J. and Löfström, J., Interpolation Spaces, Grundl. der math. Wissensch. 223, Springer-Verlag Berlin-Heidelberg-New York (1976).
3. Brudnyi, Yu. A. and Krugljak, N. Ya., Real interpolation functors, Dokl. Akad. Nauk 256 (1981), 14-17.
4. Calderón, A.P., Intermediate spaces and interpolation, the complex method, Studia Math. 24 (1964), 113-190.
5. Cigler, J., Losert, V., and Michor, P., Banach Modules and Functors on Categories of Banach Spaces, Lecture Notes in Pure and Appl. Math. 46, Marcel Dekker, New York - Basel (1979).
6. Cwikel, M., Complex interpolation spaces, a discrete definition and reiteration, Ind. J. Math. 27 (1978), 1005-1009.
7. Diestel, J. and Uhl, J.J., Vector Measures, Math. Surveys 15, A.M.S., Providence (1977).
8. Freyd, P., Abelian Categories, Harper and Row, New York-Evanston-London (1964).
9. Fuks, D.B., On duality in homotopy theory, Sov. Math. 2 (1961), 1575-1578.
10. Herz, C. and Pelletier, J. Wick, Dual functors and integral operators in the category of Banach spaces, J. Pure Appl. Alg. 8 (1976), 5-22.
11. Janson, S., Minimal and maximal methods of interpolation, J. Funct. Anal. 44 (1981), 50-73.
12. Kaijser, S., On Banach modules I, Proc. Camb. Phil. Soc. 90 (1981), 423-444.
13. Kaijser, S. and Pelletier, J. Wick, A categorical framework for interpolation theory, Lecture Notes in Math. 962, Springer-Verlag (1982), 145-152.
14. Kaijser, S. and Pelletier, J. Wick, Interpolation theory and duality, Lecture Notes in Math. 1070, Springer-Verlag (1984), 152-168.
15. Kreǐn, S.G., Petunin, Ju. I., and Semenov, E.M., Interpolation of Linear Operators, Trans. of Math. Monographs 54, A.M.S., Providence (1982).

16. Linton, F.E.J., Autonomous categories and duality of functors, J. Alg. 2 (1965), 315-349.

17. Lions, J.L. and Peetre, J., Sur une classe d'espaces d'interpolation, Inst. Hautes Etudes Sci. Publ. Math. 19 (1964), 5-68.

18. Mac Lane, S., Categories for the Working Mathematician, Springer-Verlag, Berlin-Heidelberg-New York (1971).

19. Mityagin, B.S. and Švarc, A.S., Functors in categories of Banach spaces, Russ. Math. Surv. 19 (1964), 65-127.

20. Peetre, J., H^∞ and complex interpolation, Technical Report, University of Lund (1981).

21. Peetre, J., Recent progress in real interpolation spaces, Methods of Functional Analysis and Theory of Elliptic Equations, Liquori, Naples (1983), 231-263.

22. Pelletier, J. Wick, Tensor norms and operators in the category of Banach spaces, Int. Eqns. and Op. Th. 5 (1982), 85-113.

23. Pelletier, J. Wick, Applications of the dual functor in Banach spaces, Contemporary Math. 30, A.M.S. (1984), 277-307.

24. Singer, I., Bases in Banach Spaces I, Grundl. der math. Wissensch. 154, Springer-Verlag, New York-Heidelberg-Berlin (1970).

LIST OF SPECIAL SYMBOLS AND ABBREVIATIONS

INDEX